D0024506

Venomous Snakes
of the World

Venomous Snakes
of the World

Mark O'Shea

Princeton University Press
Princeton and Oxford

Published in the United States and Canada by Princeton University Press,
41 William Street, Princeton, New Jersey 08540

First published in 2005 by New Holland Publishers (UK) Ltd
London • Cape Town • Sydney • Auckland

Library of Congress Control Number 2005920576

ISBN 0-691-12436-1

nathist.princeton.edu

Publishing Manager: Jo Hemmings
Senior Editor: Charlotte Judet
Jacket Design and Design: Gülen Shevki-Taylor
Map: William Smuts
Production: Joan Woodroffe

Reproduction by Pica Digital Pte Ltd, Singapore
Printed and bound in Singapore by Tien Wah Press (Pte) Ltd

10 9 8 7 6 5 4 3 2 1

Pictures: Page 2, Forest cobra *Naja melanoleuca*; Page 5, White-lipped pitviper *Cryptelytrops albolabris*; Page 6, MacMahon's viper *Eristocophis macmahoni* (left), Wagler's temple viper *Tropidolaemus wagleri* (right); Page 7, Eyelash palm-pitviper *Bothriechis schlegelii* (left), King brownsnake *Pseudechis australis* (right); Page 8, Rhinoceros viper *Bitis nasicornis*.

This book is dedicated to Haje, my favourite Egyptian cobra, who passed away in 2005 after 12 years as my TV co-star on numerous projects in the UK. He and I appeared in Discovery Channel's 'Ultimate Guide to Snakes', as a snake charmer and cobra team.

FOREWORD

Although Mark O'Shea is probably best known for his televised reptile-hunting adventures, his long associations and collaborations with such bodies as the Royal Geographical Society; Christensen Research Institute, Madang, Papua New Guinea (PNG); Liverpool School of Tropical Medicine and my own Centre for Tropical Medicine in Oxford, attest to his serious scientific aims. His painstaking field studies and snake collecting was the essential basis for the Oxford-University of Papua New Guinea's productive collaborative research programme on snake bites in the late 1980s and early 1990s. This work entailed detective work in pursuit of the elusive and notorious 'Papuan black snake' (*Pseudechis papuanus*) and the New Guinea small-eyed snake (*Micropechis ikaheka*) and yielded results that had an important impact on the treatment of snake bite victims throughout the country.

Mark has co-authored the successful Dorling Kindersley *Reptiles and Amphibians* as well as scientific papers and technical booklets. He now focuses on venomous snakes which are, for me, the most interesting of all reptiles, combining beauty of form and motion with a frisson of danger and enormous scientific interest. His definition of 'venomous' is acceptably pragmatic, from a human point of view (those snakes which have proved capable of causing severe envenoming). However, increasing numbers of reptilian taxa have been discovered to possess venom glands (see Underwood G. An overview of venomous snake evolution in 'Venomous snakes: Ecology, Evolution and Snakebite'. Eds. Thorpe R. S. *et al.*, Clarendon Press, Oxford, 1997: 1-13), while the venomous secretions of many species, formerly perceived as harmless, harbour potentially dangerous toxins (see Fry BG *et al.*, Isolation of a neurotoxin (alpha-colubritoxin) from a nonvenomous colubrid: evidence for early origin of venom in snakes. J Mol Evol. 2003; 57:446-52).

The introductory chapters provide useful background information on anatomy, physiology, taxonomy, venom properties, snake bites, the crisis in antivenom production and snake adaptation and ecology. The book discusses and illustrates many of the most beautiful, important and unusual venomous snakes, region by region, including one of my favourite species, the rare Central African burrowing cobra (*Paranaja multifasciata duttoni*) and the species of the greatest medical importance such as puff adders, saw-scaled vipers and spitting cobras in Africa; cobras, kraits and Russell's viper in Asia; rattlesnakes and lanceheads in Latin America and the spectacularly dangerous elapids of Australia and New Guinea.

One of the most scintillating experiences available on our planet is to observe, in an exotic location, a large, vigorous and highly dangerous serpent in the wild. Recently, I have been privileged to see wild black and green mambas in the tree tops in coastal Kenya, thanks to BioKen Snake Safaris, Watamu; and one of the Caribbean's two endemic thanatophidia, *Bothrops caribbaeus,* while wading up a river in St Lucia, courtesy of the Island's Forestry Department. Reading Mark's new book will transport you close to these delights, through his expressive field notes taken during those enviably numerous expeditions to the world's most desirable tropical locations, and through the multitude of extraordinary and original photographs, many taken by the author himself.

Conservation of these marvellous, but still profoundly unpopular, animals is one of Mark's main motivations and aims in life. The health hazards, to children, agricultural workers and hunters, posed by venomous snakes in tropical developing countries are very real, and it is perhaps understandable that, in countries where snake bite is endemic, snakes and snake-like creatures are usually killed on sight. This attitude will persist until local people can be convinced of the crucial ecological role of snakes in controlling rodents and other pests. They must be reassured that the risk of bites can be reduced through education and that snake bite victims will receive effective first aid and medical treatment. Outstanding and enjoyable books, such as O'Shea's *Venomous Snakes of the World*, will help to engage our interest and sympathy in the survival of a much-maligned suborder of the animal kingdom.

David A. Warrell,
Oxford, January 2005.

INTRODUCTION

The usual approach is to deal with venomous snake families in order: rear-fanged colubrid snakes; burrowing asps; cobras and their relatives; vipers and pitvipers. However, recent research suggests that this approach neither reflects the relationships between the families, nor their evolutionary history, the seemingly highly advanced vipers being an earlier divergence from the colubrid-elapid lineage than formerly believed. In this book we have attempted a more challenging geographical approach by dividing the world into five terrestrial and one marine chapters. Each chapter contains representatives of all the venomous snake families that occur within its geographical area.

This approach was not without difficulties – the natural world is not readily 'pigeon-holeable'. Several genera occur in more than one chapter: the cobras, *Naja* and the vipers, *Echis*, *Cerastes*, *Daboia*, *Macrovipera*, *Vipera* and *Gloydius*. It is hoped that spreading these large genera over two, in the case of *Echis* and *Naja* three, chapters does not lead to confusion.

It was also necessary to determine what constituted a venomous snake. It would have been simple to confine 'venomous' to the 300 species of front-fanged elapids and vipers, and the 18 Afro-Arabian burrowing asps. A problem of what is and what is not venomous exists in the cosmopolitan Colubridae which contains 1,700 of the known 2,800 snake species. Colubrids range from the nonvenomous American cornsnake to the dangerous rear-fanged African boomslang. Few colubrids cause serious or fatal snakebites but the majority are technically venomous as they possess venom glands or toxin-producing Duvernoy's glands. They are harmless only because they are small, secretive, or lack the ability to inject 'venom' into man. To include 'technically venomous' American gartersnakes and European grass snakes would expand the scope of the book to cover an unwieldy 2,000+ species and detract from the truly dangerous cobras and vipers . It was decided to limit colubrid inclusion to a handful of dangerous or potentially dangerous species.

Right: *All six rows of teeth are visible in the rear-fanged Blanding's Treesnake (*Toxicodryas blandingi*), two in the lower jaws, four in the upper jaws.*

Anatomy of Snakes

Snakes, together with lizards and legless amphisbaenians, belong to the order Squamata. Snakes can be summarized as elongate, scaled reptiles possessing a forked tongue and lacking eyelids, external ear openings and, usually, limbs.

External Integument

Snake scales are extremely variable in appearance and arrangement. They may be large and leaf-like, small and granular, even hexagonal in some seasnakes. Their appearance may be shiny or matt, smooth, keeled (ridged and rough to the touch), or even tuberculate (knobbly projections) in the nonvenomous filesnakes. Scales on the body are arranged in regular longitudinal rows that either overlap the scales behind, to permit locomotion in terrestrial snakes, or do not overlap, as in true seasnakes, limiting the hiding places for marine skin parasites but also preventing terrestrial locomotion. While dorsal and lateral scales are usually similar in appearance, some snakes demonstrate adaptations such as an enlarged row of scales along the vertebral apex of the back which increases a treesnake's ability to bridge gaps, or serrated, keeled lateral scales arranged in oblique rows to provide carpet or saw-scale vipers with an auditory warning to replace hissing which loses moisture in a arid environment. Ventral or belly scales are usually broad and strongly overlapping, to permit terrestrial locomotion and enhance climbing abilities, but they may be greatly reduced or even absent in burrowing or marine snakes. Some burrowers possess a sharp spine on the tail tip to force themselves through the soil or along smooth-walled termite galleries.

The scales of the head may comprise the regularly arranged, individually named scutes, of the typical 'nine-plate arrangement' or be completely fragmented into numerous tiny scales, with both extremes and every state in between represented in most of the larger snake families. Scale type, arrangement and numbers, in particular rows, are so regular and characteristic of particular snake species and genera, that they are the single most important factor in determining snake identification under field conditions. Scale arrangement and number is far more important than variable or relative characteristics, such as colour or pattern.

Coloration and patterning in snakes can be very variable with some species relying on highly cryptic camouflage, while others use permanent bright warning patterns (ie, American coralsnakes) or temporary flashes of colour (ie, boomslang, some Asian kraits and coralsnakes) to deter interference. The patterning of some harmless species may resemble that of dangerous species thereby affording them some protection through mimicry. Even within a single species colour and pattern may vary greatly, that is, sexual dichromatism where the sexes are different colours (i.e. European vipers) or ontogeneric colour changes as juveniles mature into adults. Populations, clutches or litters may even contain unicolour and striped specimens of the same species.

Below: *Thai cobra (*Naja kaouthia*), with regular colubrine-elapine nine-plate arrangement of dorsal dead scutes. Also visible is the chemosensory forked tongue and eye with round pupil.*

In place of eyelids, snakes possess transparent eye coverings known as 'brilles', which are periodically sloughed along with the rest of the upper layer of skin. The replacement of eyelids with brilles is a common characteristic of both snakes and burrowing lizards and provides protection for otherwise vulnerable eyes since moveable eyelids may have collected dirt and damaged the eyeball. The resultant unblinking gaze of snakes led to the spurious belief that they 'hypnotize' their prey and various associated stories connected more specifically with king cobras.

Skin shedding or sloughing is a regular process for snakes, related to growth, healing of wounds or the onset of oviposition (egg-laying) or birthing. Snakes may slough their skins in one entire piece, unlike lizards that tend to slough in pieces due to the presence of their limbs. Shed skins are inverted and colourless with only a faint representation of the pattern. It is not a commonly known fact but snakes even shed the skin from their tongues. This sensitive organ slides in and out of the mouth thousands of time a day and must suffer wear and tear in the process. It is therefore obvious that the outer layer must be sloughed periodically to maintain sensitivity. I have found this particularly the case with cobras.

Internal organs

Snakes are vertebrates with a skull and jaws, and a backbone to which are anchored numerous ribs. They lack a breastbone, the pectoral girdle and forelimbs, and although primitive snakes such as boas and pythons retain a vestigial pelvic girdle and remnant hindlimbs, these are absent from all venomous snakes. The organs present are the same as in any other land vertebrate, albeit modified for a tubular body. The thoracic organs include a three-chambered heart (as opposed to the four-chambered heart of mammals or crocodiles), a main lung, the right, which extends for almost half the length of the body, and a much reduced left lung.

Sense organs

Eyes may be large or small, (indistinct in the burrowing blindsnakes), with round or vertically elliptical pupils. Snakes can detect motion due to the structure of their retina which is armed with rods, rather than cones, and which provides vision similar to a mammal's peripheral vision with its 'all-or-nothing response' to danger. They cannot detect colour or shape. Some modern venomous snakes, such as cobras, appear to have particularly acute vision, responding to the very slightest movements. The highly sensitive retinas of nocturnal species are protected by a vertically elliptical pupil which can close down more completely in bright sunlight than can a round pupil, thereby protecting the retina from damage.

Snakes do not possess external ears so they are 'deaf' to airborne sound as we know it. However, most sound can cause vibrations to pass through solid material and snakes are adept at detecting vibrations through the ground, via their lower jaws. The vibrations caused by footfalls may be the first sign of an approaching prey or threat that a snake registers.

All snakes possess a chemo-sensory forked tongue to enable them to locate prey and mates and navigate within their home range. The tongue is located in the front of the lower jaws, underneath the extendible airway or glottis that permits continued breathing during the swallowing process. The forked tongue functions in collaboration with the twin-tubular Jacobson's organ in the roof of the snake's mouth. When faced with a threat a snake, venomous or otherwise, will flick its

Below: *Western Diamondback rattlesnake (*Crotalus atrox*), with typical viperine head scale granulation. Also visible is the heat-sensitive pit and eye with vertically elliptical pupil.*

tongue more frequently and more rapidly, in an attempt to determine the identity of the threat. The snake's tongue is not dangerous, contact with it does not result in the victim being 'poisoned', though close proximity to the tongue also means close proximity to the fangs and is therefore indirectly a dangerous place to be if the snake is venomous.

A few snakes, such as boas, pythons and pitvipers, also possess heat-receptors on their faces. These receptors, or 'pits', contain highly thermo-sensitive membranes, which enable those snakes to locate their warm-blooded prey. In boas and pythons the pits consist of a series of 'slit-like' openings in the lip scales of the mouth, but in pitvipers they are present as a pair of enlarged nostril-like orifices between the eye and the actual nostril. The pits face forwards, whereas the nostrils usually face backwards, upwards or sideways, and they provide the snake with instant information as to the location of the prey – both distance and direction – making an accurate lightning strike possible.

Jaws, teeth and fangs

The skull and jaws of most snakes are very flexible, they demonstrate what is known as the highest degree of 'cranial kinesis' of any terrestrial vertebrate. The reasons for this are that not only do they lack forelimbs for handling prey but they also lack specialized teeth for reducing the prey animal to smaller, swallowable pieces. Snakes must capture, restrain and swallow often large prey with what they have, a flexible skull and jaws, simple fairly unmodified teeth and either powerful constricting coils or venom. A snake's lower jaw is not united as a single unit, as in mammals, but it is, instead, comprised of a right and left half that can move independently because the lower jaw bones are not fused together to form a chin. This means that most snakes, from pythons and boas onwards, can articulate their lower jaws widely to swallow large prey animals several times the width of the snake's own head.

Snakes do not 'dislocate' their jaws, a process that would render them useless, they articulate them on the ball and socket joint where the jaw is hinged on the skull. Each side of the lower jaw is also made up of several separate and mobile bones, thereby permitting considerably flexibility in most planes. Not only can the lower jaws expand outwards to allow large prey to enter an otherwise small mouth, they also enable the snake to adopt a 'chewing motion', in combination with recurved teeth, to draw prey into the throat or advance the upper jaw and skull and bring the posteriorly positioned fangs of a rear-fanged snake into prey. The kinesis of the lower jaw replaces the need for other appendages such as hands in prey handling.

Most modern snakes possess six rows of teeth, two on the dentary bones of the lower jaws, and four in the upper jaws, on the maxilla and pterygoid bones. Most teeth are simple recurved 'hooks' which prevent the prey escaping and aid in directing it towards the throat. Some snakes have enlarged teeth in the front of the jaw, to penetrate the plumage of birds, or in the rear of the jaw, to puncture inflated frogs. Early on in the evolution of modern snakes some of the enlarged rear teeth became grooved and located below salivary glands known as Duvernoy's glands, which secreted a toxic saliva that assisted in subduing the prey. This was probably the earliest form of venom. Front-fanged venomous snakes possess highly specialized fangs positioned beneath high specialized venom glands. In the cobras and their relatives these fangs are short, and relatively fixed in position to allow the successful injection of venom into the prey. In the vipers the fangs are much longer, so long that the snake could not close its mouth with them erect, so viper fangs are hinged and lie flat, horizontal with the jaw, when in the resting position. To permit this resting position, vipers possess a diastema, a toothless gap, behind the fangs and effectively only possess two rows of teeth in the upper jaw. When a viper strikes the long fangs swing forwards into the strike position and the skull demonstrates enormous cranial kinesis as it prepares for impact. At the strike the closing of the lower jaw on the prey applies further penetration by the fangs of the upper jaw. While cobras and their kin will often deliver a prolonged 'chewing bite' as the fangs are repeatedly and alternatively stabbed into the prey, a viper will usually deliver one rapid, double-barrelled, stabbing bite with both fangs, instantly releasing and sitting back to await the results of the strike. This is known as a 'stab-and-release bite' and is a common strategy applied by a 'sit-and-wait ambusher', such as a rattlesnake or puff adder, that then sits back and yawns widely to realign its fangs, before using its forked tongue to track down its strickened prey. These long fangs, attached to the highly kinetic upper jawbones, are also used to 'fang walk' the dead prey down the throat. Since they are hinged and get a great deal of use, they are frequently lost, either falling from the mouth or passing into the stomach with the prey animal. For this reason there is usually a replacement fang present on either the right or left side.

Opposite top: *Colubridae: Boomslang (*Dispholidus typus*) - the enlarged rear-fangs and highly toxic venom make this a dangerous species.*

Opposite middle left: *Atractaspididae: Reticulated burrowing asp (*Atractaspis reticulata*) - the long horizontal fangs can be stabbed backwards without the snake opening its mouth.*

Opposite middle right: *Elapidae: Coastal taipan (*Oxyuranus scutellatus*) - the fixed front fang, long for an elapid, is visible, connected to the venom gland by the venom duct.*

Opposite bottom left: *Elapidae: The yellow-lipped sea krait (*Laticauda colubrina*) has small fixed elapid fangs.*

Opposite bottom right: *Viperidae: Russell's viper (*Daboia russelii*) - the long, hinged viperid front fang is visible followed by a toothless diastema with the venom duct and venom glands above.*

Venomous Snake Diversity and Distribution

Suborder Ophidia (or Serpentes) comprises approximately 2,800 living snakes species in 16 to 20 families, split between three superfamilies. All venomous snakes are 'advanced snakes', members of the superfamily Colubroidea, or Caenophidia. They are found in most habitats where snakes occur, and are responsible for upwards of 40,000 annual human fatalities worldwide.

Evolution of venomous snakes

The oldest known fossil snake comes from the Cretaceous period 95 million years ago, but it is possible that snake-like reptiles existed in the Jurassic period, 140 million years ago. Exactly what was the ancestor of the snakes is open to argument, but an ancestral varanoid (monitor-type) lizard or an, even earlier, elongate anguimorph (slow worm-like) lizard are possible candidates, with extinct paddle-limbed marine mosasaurs suggested as another possibility. The earliest true snakes inhabited the southern super-continent of Gondwanaland and were small terrestrial or burrowing animals, probably not unlike modern blindsnakes or pipesnakes, limbless with elongated bodies and vestigial eyes. They were, of course, nonvenomous.

Larger snakes, the ancestors of modern boas and pythons, with large appetites and mouths to match, evolved by the Eocene epoch (55 million years ago), but even though some of these serpents achieved 'anaconda rivalling' lengths of 5 to 10 metres, before largely disappearing by the Oligocene epoch (35 million years ago), they too were nonvenomous.

Venom came with the later colubroid snakes which first made an appearance in the Miocene epoch (25 million years ago). There was a sudden explosion in diversity within the colubroid snakes in the northern super-continent of Laurasia during this time, possibly as a reaction to a similar explosion in the evolution and availability of small mammals and other prey species and the opening up of numerous suitable habitats and niches. So considerable was the rapid diversification that this period has been called 'The Age of the Snakes'. It was a commonly held belief that 'advanced' venomous snakes, such as cobras and vipers, evolved much later from within the nonvenomous colubrids, possibly by way of a rear-fanged venomous colubrid or the atractaspid burrowing 'mole vipers', but this idea is outdated, especially since fossil records for ancestral cobras and vipers are almost as old as fossil records for colubrid snakes. Fossil colubrids and vipers are known from mid-Miocene Europe and late-Miocene North America, while the earliest cobra fossils are from early Miocene France and Spain.

Diversity of modern snakes

A breakdown of Colubroid diversity:

Order: Squamata (scaled reptiles)
Suborder: Ophidia (snakes)
Superfamily: Colubroidea (advanced snakes)
Family: Acrochordidae – 3 spp., Indo-Australia, aquatic and nonvenomous.
Family: Colubridae – c.1700 spp., worldwide, all habitats, nonvenomous and rear-fanged venomous.
Family: Atractaspididae – c.30 spp., Afro-Arabia, terrestrial, venomous.
Family: Elapidae – c.315 spp., worldwide excluding Europe, all habitats including marine, front-fanged venomous.
Family: Viperidae – c.270 spp., worldwide excluding Australasia, all terrestrial habitats, front-fanged venomous.

Family Acrochordidae

The two freshwater and one marine filesnakes of Indo-Australia are nonvenomous and pose no danger to humans.

Family Colubridae

This is the largest and most diverse family and also the most widely distributed, with representatives almost everywhere that snakes exist. There are around 1700 species in the Colubridae, distributed between up to 12 subfamilies, ranging from harmless nonvenomous snakes such as the American cornsnake to dangerous rear-fanged venomous snakes such as the African boomslang. There are no front-fanged colubrids but although the number of species that have caused human fatalities can be counted on the fingers of one hand, that number is in danger of climbing to two hands as accidents with obscure colubrid species increase in number. Most species are technically venomous, since they possess Duvernoy's glands and several species that were previously considered harmless have now been found to possess powerful neurotoxins known as '3-fingered toxins', which may rate alongside the neurotoxins of cobras in their toxicity. However most colubrids lack the delivery mechanism for injecting the toxins into humans and it is not known if the venoms would have any effect even if they were administered to a human. It is now believed that only the European and American ratsnakes, American kingsnakes and a few closely related colubrids can be truly considered nonvenomous, although most others can probably still be considered harmless. Rear-fanged species were sometimes called 'opisthoglyphous', which means 'knife behind' and refers to the enlarged grooved fangs positioned at the rear of the jaw, while species lacking enlarged rear-fangs were known as 'aglyphous' or 'without knife'. Those species that have caused serious snakebites are contained in the largest three colubrid subfamilies, the cosmopolitan Colubrinae and Natricinae, and the American Xenodontinae. Some authors now elevate the freshwater keelback subfamily to full familial status as Natricidae.

Family Atractaspididae

This is an Afro-Arabian group which had been shuffled around between the Colubridae and the Viperidae, but which are now recognized as a separate, distinct family in its own right. Containing approximately 18 species of mole vipers, burrowing asps or stiletto snakes – the side-stabbing snakes – the Atractaspididae is usually expanded to contain another dozen or so related African snakes, such as the Natal balcksnake and the centipede-eaters. The burrowing asps have long, horizontal fangs and can bite without opening their mouths, flicking the fangs out and backwards as they jerk sideways. They are capable of inflicting painful snakebites, although fortunately only three species are attributed with causing human fatalities.

Family Elapidae

The Elapidae contains approximately 315 species formerly known as the Proteroglyphs, or 'early-knived' snakes, a reference to the positioning of their short, fixed fangs at the front of the upper jaw. This is a very diverse family that is found in the Americas, Africa, the Middle East, tropical Asia to Australasia and the Indian and Pacific oceans. It was previously divided into several subfamilies but is currently recognized as containing just two: the Elapinae, terrestrial and freshwater snakes from the Americas, Africa and Asia (coralsnakes, cobras, mambas, kraits etc.) and the Hydrophiinae containing all venomous terrestrial Australo-Melanesian snakes (except possibly one species from Bougainville) and the marine seasnakes and sea kraits.

The Elapidae contains some of the smallest front-fanged venomous snakes, such as the small, secretive Papuan forest snakes and the harlequin snakes of Africa, and also most of the long slender species which reach far greater lengths than achieved by the stout-bodied vipers, such as the black mamba and taipan, which grow to 3.5m and the king cobra, which grows up to 6m. Most elapids resemble colubrid snakes, from whom they probably evolved, the head is rounded and of similar width to the neck, and the head and body scales are also very similar. However, the Australasian death adders look much more like vipers than colubrids or elapids, having evolved to occupy the vacant niche for a sit-and-wait ambusher in a land without vipers. The scales of elapids are smooth, only a few elapids possessing keeled scales,

Above: *The Common lancehead (*Bothrops atrox*) is the number one cause of serious snakebites in Amazonian South America. Its long, hinged fangs deliver venom deeply into the tissues.*

DIVERSITY OF COLUBRIDAE

Subfamily	Species causing fatalities	Distribution	Genera causing serious snakebites	Distribution
Colubrinae	*Dispholidus typus* *Thelotornis* spp.	Africa	*Toxicodryas* *Boiga*	Africa, Asia & Australasia
Natricinae	*Rhabdophis tigrinus*	Japan	*Rhabdophis*	South East Asia
Xendontinae	*Philodryas olfersii*	Brazil	*Phalotris* *Tachymenis* *Hydrodynastes* *Philodryas* *Xenodon*	South America

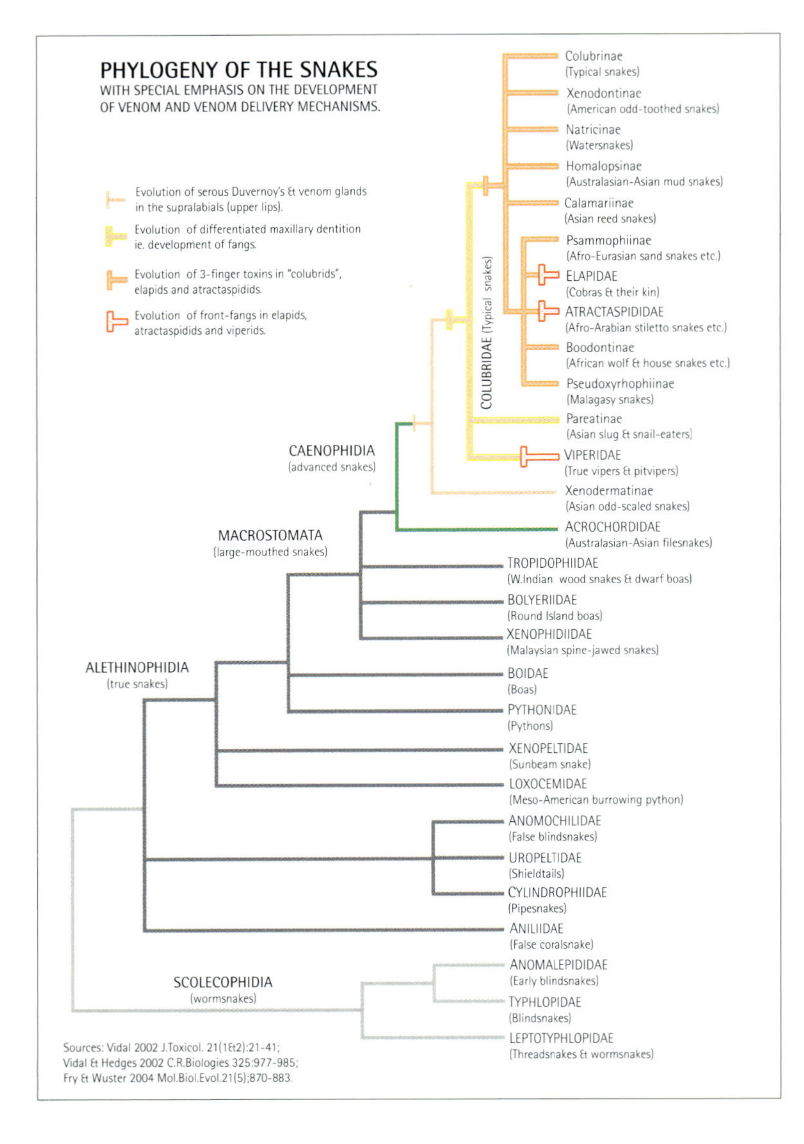

Above: *The phylogeny or origins of snake families is a constantly disputed subject. The above chart combines the results from several recent papers that demonstrate that the evolution of venom and venomous snakes is not as recent as previously assumed.*

for example, the Australian rough-scaled snake, the African ringhals and, to a lesser degree, some death adders. Although many of the small secretive elapids have tiny eyes, the highly alert diurnal species, for example, the mambas and tree cobras, have large eyes and acute vision. Most species, but not all, have round pupils. While many elapids are patterned with subdued browns and camouflaged greens, this family does contain some of the most startlingly patterned of all snakes, such as the brightly tricolour-banded American coralsnakes and red-bellied long-glanded Asian coralsnakes.

Defensive techniques vary from cryptic patterning and warning coloration, to hooding and spitting venom in the cobras. Most elapids are oviparous (egg-laying) but southern species, like the South African ringhals and Australian tigersnakes, are viviparous (live-bearers). When eggs are laid they are usually left to incubate on their own but the female king cobra will build a nest and guard it until the eggs have hatched.

Family Viperidae

The Viperidae contains the snakes formerly known as the Solenoglyphs, or 'pipe knived' snakes, a reference to their hollow fangs, as opposed to the fangs of the elapids which are not fully enclosed and are more caniculate. Four subfamilies are recognized: the monotypic primitive Azemiopinae from Myanmar and Vietnam; the primitive colubrid-like Causinae from Africa; the Viperinae, true vipers of Eurasia and Africa; and the Crotalinae, pitvipers from the Americas and Eurasia.

Many vipers are terrestrial, of a moderate to stout build and range in size from tiny 0.3m Sri Lankan carpet vipers to 2m African gaboon vipers. Many are sedentary, sit-and-wait ambush predators of ground-dwelling rodents, lizards and frogs. Other species are agile and highly arboreal with fairly slender bodies and prehensile tails, such as the eyelash palm-pitviper or bushvipers, which actively hunt treefrogs, birds or bats. A few, such as the cottonmouth, are found in freshwater habitats pursuing frogs and even fish. There are approximately 270 species of vipers distributed throughout the world, excluding Australasia, the oceanic islands and, of course, polar regions.

All vipers possess a pair of long, hollow, hinged fangs in the front of the upper jaw. At rest they lie parallel with the upper jaw, as they are too long to permit mouth closure when erect, but when the viper strikes they move forward into a position

DIVERSITY OF ELAPIDAE

Subfamily	Content	No. of genera & species		Distribution
Elapinae	Cobras, mambas & coralsnakes	19	145	Africa, Asia, Americas & Bougainville
Hydrophiinae	Australasian & marine elapids	47	170	Australasia, Melanesia & Oceania
		66	315	

DIVERSITY OF VIPERIDAE

Subfamily	Content	No. of genera & species		Distribution
Azemopinae	Fea's viper	1	1	N. Myanmar, N. Vietnam & S. W. China
Causinae	Night adders	1	6	Sub-Saharan Africa
Crotalinae	Pitvipers	23	183	Eurasia & the Americas
Viperinae	True vipers	12	80	Europe, Asia & Africa
		37	270	

where they will contact the prey or victim and be driven deep into the tissue Because of the need to rest the fangs in the horizontal position, vipers lack other teeth from the outer, upper jaws and instead possess a gap known as a diastema. The presence of a pair of heat-sensitive pits enables the pitvipers to accurately locate warm-blooded prey, even in the total dark of the jungle night, and this provides them with a massive advantage when hunting. Most vipers are cryptically patterned, they rely on camouflage to blend into their surroundings, an essential part of being a sit-and-wait ambusher, allowing the prey to come within strike range. Some cryptic vipers have brightly coloured tail tips that they wave to attract hungry prey within range of a strike, a method known as 'caudal luring'. Although some arboreal vipers are brightly coloured golden yellow or even white, this may too be a form of camouflage because prey may mistake a coiled, motionless yellow viper for fruit, leaves or flowers.

Most vipers have vertically elliptical cat-like pupils that maximize the use of available light in the low-light jungle or nocturnal conditions but protect the sensitive retina in bright light by closing down more efficiently than a round pupil, but the primitive night adders have round pupils like most colubrids. The primitive dorsal head-scale condition is the colubrine 'nine-plate arrangement', where the top of the head is covered by nine large well-defined plates or scutes: (from front to back) 2 internasals; 2 prefrontals; 1 frontal; 2 supraoculars; and 2 parietals. This pattern demonstrates the vipers ancestral link with the colubrid snakes and is clearly visible in more primitive members of the Viperidae, such as Fea's viper in Asia, the night adders in Africa and the copperhead and pigmy rattlers among the Americans pitvipers.

The more advanced vipers show increasing degrees of head-scale fragmentation to the extremes of the bush vipers and gaboon viper of Africa and diamondback rattlesnakes of America. In between there are species that demonstrate the part-way stages in the evolution towards scale fragmentation with a mixture of scutes and small scales, for example, some of the rare montane rattlesnakes of Mexico and the true vipers of Eurasia.

Most vipers are viviparous, which is the secret to their successful colonization of cool-temperate habitats. In cold conditions abandoned nests of eggs would perish but a female carrying her offspring internally until full term can seek out sunspots and avoid cold spells to increase their chance of being born and surviving. This is why the northern-most snake in the world, the European cross adder, can live within the Arctic Circle of Scandinavia and the southern-most snake, the Patagonian lancehead, a pitviper, can inhabit southern Patagonia. Not all vipers are viviparous, some tropical species do lay eggs, notably the bushmasters of South America, the Malayan pitviper and some sand and carpet vipers in Afro-Arabia.

Many vipers hiss loudly in defence, for example, the puff adder gets its common name from the loud noise it makes when disturbed, but other vipers inhabiting arid conditions, where repeated hissing would result in considerable body moisture loss, have developed unique methods of issuing a warning. Carpet or saw-scale vipers rub their keeled and serrated lateral body scales together to make a loud sawing sound while rattlesnakes have evolved hollow rattles as a warning device. Every time the rattlesnake sheds its skin it adds one link to the proximal end of the rattle. The rattle sections are interlinked and the characteristic sound is produced when they are vibrated against one another, powered by muscle in the tail that never fatigues and can continue functioning indefinitely (similar to cardiac muscle). Contrary to common belief it is not possible to age a rattler by the number of links in its tail because although it adds one link each time it sloughs its skin, that may not be at a regular rate each year and most wild rattlesnakes possess rattles with the distal links broken or missing. Some rattlesnakes, which live on islands that lack natural enemies, have undergone the process of losing their rattles since they are more of an auditory encumbrance when hunting birds roosting in low bushes.

Snake Venoms and their Actions

There are many types of snake venoms and although they may be grouped into those that affect the nervous system, those that affect the blood and those that affect the tissues, it is not as simple as that because there are numerous different toxins in each group and some toxins may have more than one effect. What is more, snakes often contain combinations of toxins in different concentrations, which vary with the age, diet and distribution of the snake species and cause differing effects in prey to their effects in humans. Comparing venoms is also a difficult problem. When one is asked which is the most dangerous, a rattlesnake or a cobra, you are comparing two very different, and in themselves, variable venoms, and it is much like comparing apples with lemons or bananas. They are all fruit, but of different shapes, textures and tastes. With the fruit you might choose to compare their 'sweetness' because they all contains sugars, albeit different types of sugars.

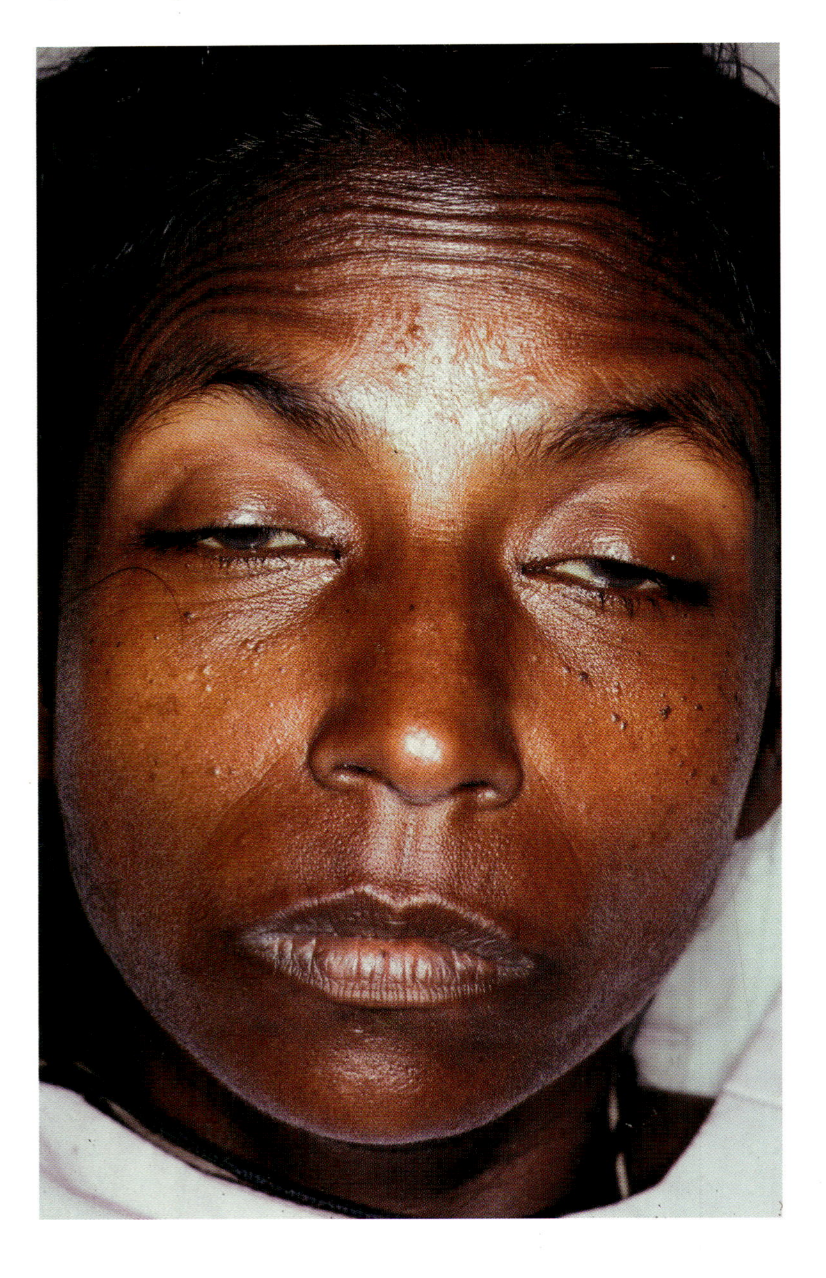

A similar process has been set up for venoms called the LD_{50} test, whereby venoms are compared by their 'killing capacity'. LD_{50} stands for 'median lethal dose' in other words the amount of venom that will kill 50 per cent of a test group of 20gm mice. Clearly, the lower the figure, the more toxic the venom (to mice). In this way different venoms that attack nerves, blood, tissues etc. can be compared based on their killing capacity in mice, and so far the lowest doses recorded, and therefore the most venomous snakes known, are all Australasian, the inland taipan (*Oxyuranus microlepidota*) with 0.010mg/kg, the eastern brownsnake (*Pseudonaja textilis*) with 0.040mg/kg, and the coastal taipan (*O. scutallatus*) at 0.064mg/kg. When considered alongside average venom yields for these three snakes – 44mg, 2mg and 120mg respectively – it can be that each species is theoretically capable of killing 220,000, 2,500 and 93,750 20gm mice with a single bite.

There are two main problems with LD_{50} results, apart from the obvious ethical issues. The first is that the method of injection may vary and result in hugely different figures. LD_{50} tests are usually done with *im* (intramuscular) injections although they may also be given *iv* (intravenously), *sc* (subcutaneously – under the skin) or *ip* (intrapenitoneally – into the body cavity). Of the above, all, except *iv*, are likely routes of injection when a snake strikes small prey but in humans, a defensive bite is much more likely to be *sc* or *im*, into the body tissues. The other problem is how to determine lethal human doses from lethal mouse doses. In the past this has been done by simply extrapolating the result from a 20gm mouse up to a 70kg human. Clearly this is not ideal because people weigh different amounts and species react in different ways to different venoms. A venom that has a neurotoxic effect on mice does not necessarily have the same effect on a human. It is now possible to determine how much venom has been injected into a victim by various analytical techniques that look for serum levels etc.

NEUROTOXINS:

Action: Neuromuscular paralysis varying in severity from minor to fatal.

Early signs: 1. Pre-paralytic: dizziness, headache, loss of senses of taste and smell but increased sensitivity to sound, and goosebumps. 2. Paralytic: ptosis (drooping eyelids), ophthalmoplegia (double vision), flaccid facial paralysis (inability to smile, close mouth and eyes tightly), paralysis of the tongue and mouth muscles and the production of and inability to swallow increased quantities of saliva.

Left: *A common krait (*Bungarus caeruleus*) bite demonstrating the early sign of neurotoxic envenoming, 'ptosis', the paralysis of the eyelid muscles.*

Serious signs: Paralysis of the muscles used for swallowing and breathing, so that saliva accumulates in the back of the throat, with a danger of inhalation/aspiration and death occurring from asphyxia, through paralysis of the diaphragm and other muscles of respiration.

Effects: Presynaptic neurotoxins act on the proximal terminal of the nerve synapse by blocking the release of acetylcholine, the physiological transmitter and eventually destroying the nerve terminal, preventing transmission of nervous impulses across the synaptic gap. Postsynaptic neurotoxins act on the distal terminal of the nerve synapse by competitively inhibiting the binding of acetylcholine, again preventing the physiological transmission of nervous impulses across the synaptic gap. Once established, the paralytic effects of presynaptic neurotoxins cannot be reversed with antivenom. The victim must be artificially ventilated until new synaptic terminals can develop. Paralytic effects caused by postsynaptic neurotoxins are sometimes reversible with antivenom and victims may demonstrate rapid and complete recoveries. Most snake venom presynaptic neurotoxins are phospholipases A2, while postsynaptic neurotoxins are peptides with a 'three-finger' structure allowing them to bind to the acetylcholine receptor at the neuromuscular junction. Many snake venoms contain both pre- and post-synaptic neurotoxins.

Many snake venom toxins interfere with normal mechanisms of 'haemostasis' (that stop bleeding after injury) and cause bleeding through several different mechanisms:

1. Procoagulant actions, which use up clotting factors.
2. Anticoagulant actions, which inhibit normal clotting.
3. Anti-platelet actions, which inhibit normal clotting.
4. Haemorrhagic actions, which damage the lining of blood vessels.

COAGULANTS:

Action: Reduction in the blood's ability to coagulate or clot, leading to coagulopathy (incoagulable blood).

Signs: Continued bleeding from the site of the snakebite and other wounds. 'Twenty-minute whole blood-clotting test' involves placing fresh blood in a clean test tube and checking to see if it has clotted after 20 minutes. Normal blood will have clotted but the presence of coagulants or lack of platelets will prevent clotting and leave the blood liquid.

Effect: Procoagulants initially promote blood-clotting, but the body's own clot-busting system breaks down the clots as quickly as they form until the supply of clotting factor is exhausted and the blood indirectly becomes incoagulable. If clots are not broken down by the body there is a possibility for circulating blood clots to cause thrombosis, the most immediately fatal form being cerebral or pulmonary thrombosis or strokes. This happens rarely in snakebite victims, except following bites from the pitvipers of St Lucia and Martinique in the Caribbean. Anticoagulants prevent normal blood-clotting, making blood incoagulable. Some snake venoms contain both pro- and anticoagulants.

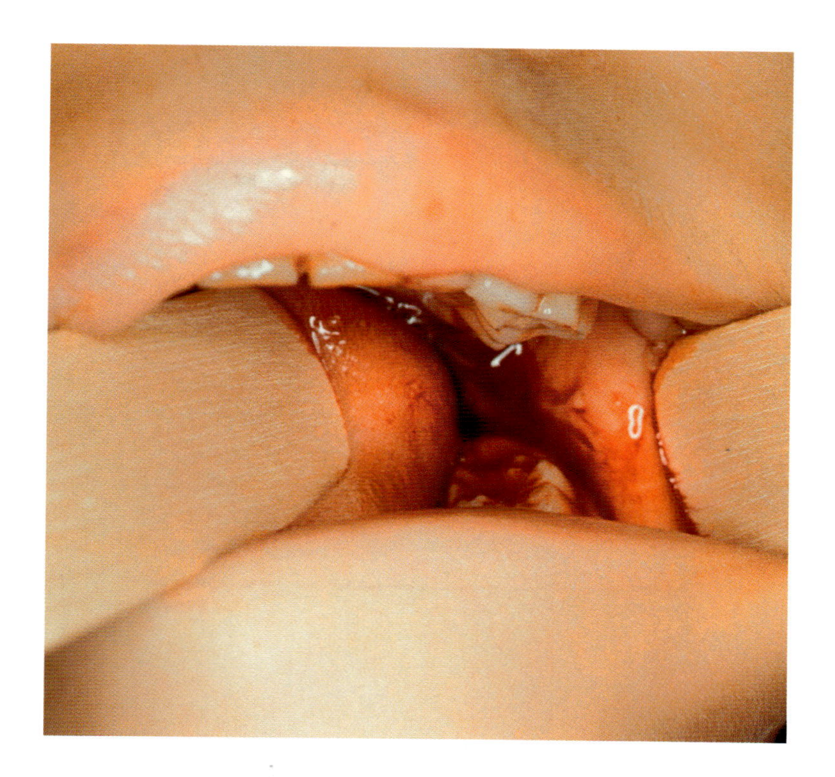

Above: *An early sign of the haemorrhagic effects of Malayan pitviper (*Calloselasma rhodostoma*) venom is bleeding from the gums, old scars, the bite-site and other orifices.*

HAEMORRHAGINS:

Action: Blood vessels become permeable to blood leakage into the body tissues.

Early signs: Bleeding from gums or nose, the site of the snakebite, old scars, 'first aid' cuts, ulcers, and blood in the sputum, saliva, vomit, urine or stools.

Serious signs: Sudden severe headache, 'stroke', shock or neck stiffness.

Effects: Combination of procoagulants and haemorrhagins can cause rapid and fatal loss of blood into the tissues leading to hypotension (lowered blood pressure) and fatal shock. Death may be caused by cerebral haemorrhage (bleeding into the brain, 'stroke'). Renal failure may result from bleeding into the kidneys. Thrombcytopenia (a reduction in the number of platelets or thrombocytes, which assist blood clotting) or venom-induced platelet functions without thrombocytopenia, may exaccerbate bleeding problems. Blood platelets congregate at any wound and clump together to prevent blood loss. Normal blood platelet counts are 150–400. Serious snakebites can cause levels to drop to 30 or as low as 5 (as in one of my rattlesnake bites) and at this level there are insufficient platelets to prevent bleeding which will continue unabated until the patient bleeds to death or suffers renal failure.

HAEMOTOXINS (also known as HAEMOLYTIC):

Action: Haemolysis (breakdown of red blood corpuscles).

Signs: Haemoglobinuria (passage of the blood pigment haemoglobin into the urine), resulting in pink urine.

Effects: High levels of haemolysis cause renal failure due to blood debris blocking the kidneys and preventing them from functioning normally.

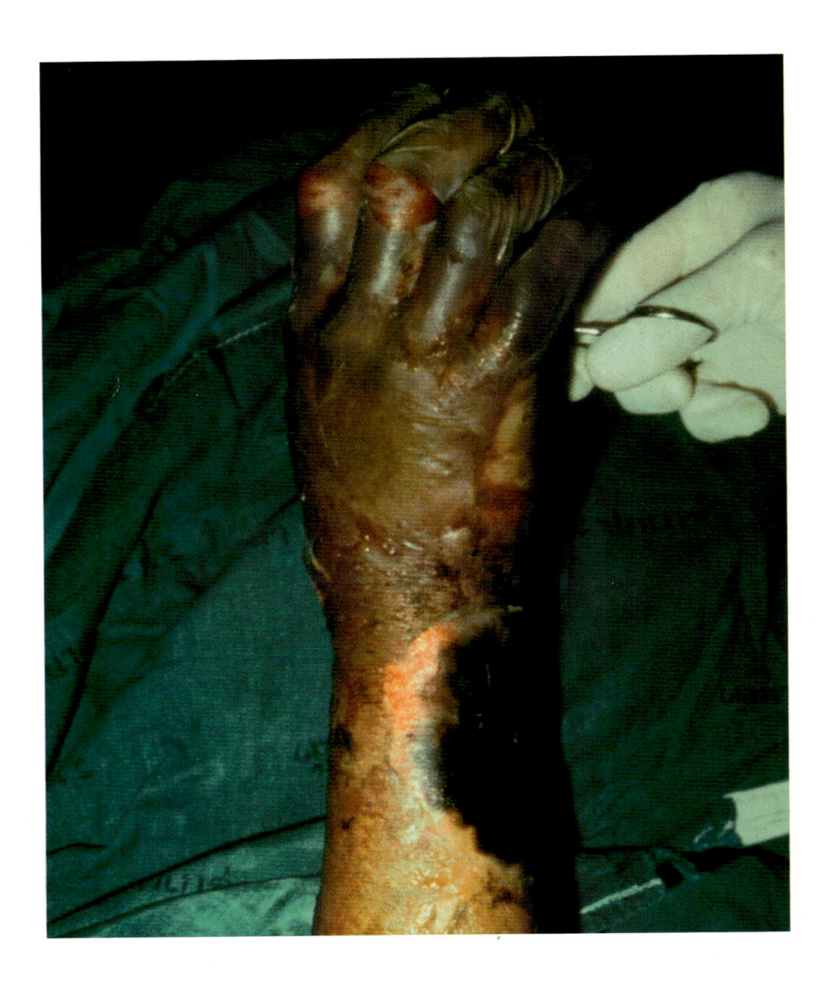

Above: *A Malayan pitviper (*Calloselasma rhodostoma*) bite causes massive destruction of tissue due to its cytotoxic properties. This bite was on the now blackened wrist.*

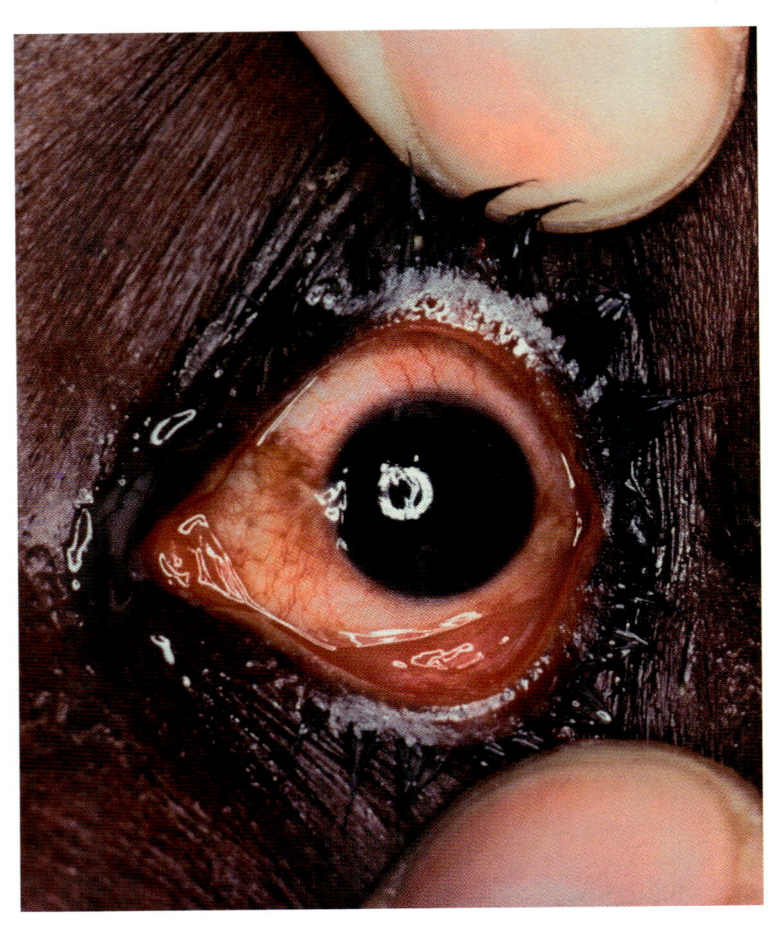

Above: *Black-necked spitting cobra (*Naja nigricollis*) venom in the eye causes extreme pain, temporary blindness and may cause permanent corneal damage and blindness.*

MYOTOXINS:

Action: Damage to muscles, especially the respiratory muscles, either by paralysis of the neuromuscular function (similar to neurotoxins) or by rhabdomyolysis (breakdown of the skeletal muscle).

Signs: Myoglobinuria (passage of muscle oxygen carrying pigment, myoglobin, into the urine), resulting in brown urine (the colour of Coca-Cola) darker than haemoglobinuria.

Effects: Death may be due to kidney failure caused by excessive myoglobin release from damaged muscles or cardiac failure due to hyperkalaemia (an increase in blood potassium caused by muscle breakdown). One of the commonest venom components causing rhabdomyolysis is phospholipase A2 similar to those causing presynaptic neurotoxic paralysis.

CYTOTOXINS (also known as NECROTOXIC or PROTEOLYTIC):

Action: Breakdown of tissues as a result of venom enzymes designed to speed up the process of digestion of prey.

Signs: Massive swelling, pain, discoloration, blistering, bruising, wound weeping.

Effects: Although not directly life-threatening, cytotoxins cause tissue necrosis (tissue-death). Necrosis may occur anywhere on the body but is commonest around the site of the bite and if that is a finger, toe, hand or foot, it may lead to gangrene and limb loss. Fasciotomy (gross surgical opening of tissue compartments in the bitten limb to relieve pressure on blood vessels and other organs), amputation, skin grafts and long periods of convalescence may be required to deal with the injury. Secondary infection with bacteria or development of cancers may be later consequences of cytotoxic snakebites.

NEPHROTOXINS:

Action: Direct damage to the kidneys.

Effects: Renal failure may be a consequence of other venom components causing direct bleeding into the kidneys, such as haemorrhagins, blockage of the kidney nephrons due to a build-up of blood and muscle pigment debris caused by haemotoxins and myotoxins, and tissue-oxygen starvation due to hypotension (lowered blood pressure), but some venoms directly target the kidneys themselves and cause renal failure. Also microclots in the renal blood vessels and the constriction of renal blood vessels may restrict the blood flow.

SARAFOTOXINS:

Action: Found only in burrowing asps of Afro-Arabia and causes coronary artery vasoconstriction (narrowing of blood vessels).

Effects: The narrowing of the coronary arteries will cause reduced blood flow and may lead to a heart attack.

For information on antivenoms see page 154.

Life in the Sea

Seasnakes and sea kraits have had to adapt in many subtle ways to become successful ocean-going predators. They are probably amongst the most environmentally adapted of all snakes. Here I will attempt to summarise the specialized adaptations of the marine snakes to their environment.

1. They have flattened paddle-shaped tails and some seasnakes, with small ventral scales, can flatten their entire bodies for effortless swimming.

2. Valves in the nostrils prevent water entering and tight fitting scales along the lips perform a similar task.

3. Water loss to the saline ocean is limited by a seasnake's skin, which is less permeable to water passing outwards than that of a terrestrial snake.

4. Excessive salt levels are highly dangerous. Sea kraits drink freshwater on land or lap rain from their scales while seasnakes can drink rainwater that pools on top of seawater after heavy storms. Salt is taken in with prey but the excess cannot be expelled through the skin or via the kidneys. Seasnakes possess salt excretory glands in mouth that transfer salt onto the tongue which carries to the outside ocean.

5. Seasnakes can dive to 100m, and remain on the bottom for up to two hours, though 30 minutes is more the norm, and return to the surface rapidly for another breath. Such feats raise numerous complicated questions relating to respiration rates, oxygen requirements, 'the bends' and buoyancy. A cold-blooded reptile uses oxygen at a slower rate than a mammal and may remain submerged for 6-7 times longer than a similar sized mammal, 15 minutes to a mammal's two minutes. Many diving reptiles still built up an anaerobic 'oxygen debt' but seasnakes are even better adapted using 'breathing tachycardia', a process by which a rapid breathe leads to efficient absorption of more oxygen into the blood in readiness for the next dive. Air is stored in a muscular saccular lung which forces it back into the vascular lung when it is required for respiration. Seasnakes also do not suffer from a Bohr shift caused by a build up of carbon dioxide, as diving terrestrial reptiles might. Their skin is 30x more permeable to gaseous exchange than other snakes' skin so cutaneous respiration is possible when pulmonary respiration is impossible.

6. A snake's heart is three-chambered with deoxygenated and oxygenated blood alternately passing through the common ventricle permitting partial mixing and re-routing the blood via a 'cardiac shunt'. A left to right shunt sends more blood to the lungs, enhancing gaseous exchange during extreme activity, while a right to left shunt sends more blood to the body tissues of a basking snake and speeds up the thermoregulation process. The shunt sending more blood to the body could benefit a submerged seasnake in that gaseous exchange through the skin would be enhanced and air in the lung could be conserved. A left to right shunt at the surface would make more blood available during for oxygen absorption, enhancing the effects of breathing tachycardia. Seasnakes also possess a blood vessel connecting the pulmonary artery to the pulmonary vein allowing blood to bypass the lungs in a 'pulmonary shunt', isolating the lungs when pulmonary respiration is not possible and cutaneous respiration is being utilised.

7. A seasnake's heart is positioned centrally in the body since water pressure maintains blood pressure throughout and central positioning equalizes the blood flow to both ends of the body.

8. Some seasnakes dive to 100m then surface at a rate of 1m/3sec., which could take 10 minutes. A mammal diving in this manner would suffer from the 'bends' as nitrogen came out of solution in the blood as the pressure reduced upon surfacing. Seasnakes seem unaffected. One argument suggests that they do suffer from the bends but they return to depth so rapidly, after surfacing for a single breath, that nitrogen bubbles go back into solution before any damage is done. This could explain the deaths of apparently healthy seasnakes, captured as they surfaced and prevented from diving again. The counter argument points out that seasnakes often lie on the surface for long periods, especially at night. Both may be true, the surface floating snake having taken longer to surface, allowing time for the nitrogen to exit its circulation. Also if seasnakes were to isolate the lung, either by a right to left cardiac shunt or a pulmonary shunt, they might reduce the risk of the bends and expel all dissolved nitrogen through the skin before surfacing. Deep diving seasnakes have a pronounced cardiac shunt that enables them to dive to 100m and back without ill effect; more shallow dwelling sea kraits have a less effective shunt, and are therefore less effective divers.

9. The final adaptation concerns air as a buoyancy aid. A seasnake at the surface with a lung full of air is positively buoyant but when it dives deep a combination of the water pressure, compacting the air in the lung, and loss of air through cutaneous respiration reduces the volume of air, making it negatively buoyant, and sink. A surface dwelling species, such as the yellow-bellied seasnake, could experience difficulty regaining the surface when negatively buoyant so it may only dive to escape rough weather. Bottom-dwelling fish-egg-eating species may ingest sand as ballast to help keep them on the bottom. Even repositioning the air in the lung may help a snake dive or surface.

Conservation

It might be thought that snakes are a dangerous risk to human life and it is true that between them, venomous species are probably responsible for the deaths of upwards of 40,000 humans per annum, although most of those fatalities will probably be caused by a short list of about thirty species. 40K is a large number certainly, but it is far fewer than the numbers killed by the malarial mosquito, or war, and the disease and famine that follows in its wake, or even by automobiles, smoking, cancer and heart disease.

No venomous snake can be said to be actually threatening the very existence of *Homo sapiens*, even though they may make certain locations fairly inhospitable and undesirable places to set up home – Ilha Queimada Grande, off São Paulo, Brazil and Chappell Island, off Tasmania, springing to mind. The reverse, sadly, cannot be said of man's affect on venomous snakes.

Snakes, including venomous species, face numerous human threats to their existence. Some years ago I wrote that snakes were usually killed or collected for one of the four F's: from Fear, for Food, for Fun, or for Financial gain. I think this probably still sums up the situation.

The most obvious threat is the indiscriminate and wanton destruction by persons encountering them in the wild, what used to be called 'adder bashing' in the United Kingdom. This is usually small scale, driven by fear and ignorance and is as likely to occur to a nonvenomous snake as a venomous species. Many people fear snakes and see it as their duty to rid the planet, or at least their small patch of it, of the accursed serpent before it has the chance to turn the tables. Small scale destruction is misguided and may even be dangerous because snakes tend not sit idly by while their lives are extinguished, they defend themselves and many people have received serious snakebites from snakes that they were killing, or thought they had killed. However, it is unlikely to have a serious affect on the overall snake population, unless it is already in dire straits. Organised mass destruction, the offering of bounties for snake's heads or rattlesnake's tails, as occurred with timber rattlers in eastern USA and tigersnakes in Australia in the early 20th century, are much more serious and threatening. It was out of this 'need' to eradicate the snake population and prevent them biting stock animals that the infamous North American Rattlesnake Round-ups emerged. Huge numbers of rattlers were collected and kept together in crowded and unhygienic conditions, ready for a great orgy of destruction and demonstration of superiority of man over serpent. The round-up methods were very questionable and as people sought to find every single venomous snake, rocky outcrops were dismantled and petrol was poured down burrows to drive out the snakes, killing many other inhabitants in their dens and doing irreparable damage to the ecology. These round-ups have now become multi-million dollar tourist attractions where people compete against the clock and each other, to stuff live rattlers into sacks, and risk bites in the process, or eat more snake meat that the next man. Round-ups do a great deal of damage, snakes are shipped in from neighbouring states to satisfy the show's needs, no longer is it a pest removal service, more a pest importation exercise which may threaten the existence of isolated populations of rare species.

Below: *This Mexican west coast rattlesnake (*Crotalus basiliscus*) and Beaded lizard (*Heloderma horridum*) were killed and thrown onto the road by a Mexican grass-cutting crew. The lizard is a protected species. Notice that the rattlesnake's rattle has been taken.*

If round-up supporters trumpet the rattlesnake as a biological threat others see it in a more sinister light, as the implement of the devil. In the Appalachian Mountains of West Virginia is to be found the Church in Lord Jesus where the parishioners base their entire philosophy of five lines from Mark [XVI:18], including "They shall take up serpents" which they do literally, vanquishing the devil by handling venomous snakes during the services. If bitten they refuse treatment and either survive or succumb, trusting in The Lord at all times. They might see snakes as evil but do not threaten their existence like the round-ups a little further west.

Snakes, especially venomous species, have found their way into the psyche and customs of almost every culture they encountered. Hopi indians in Arizona used to perform a ceremonial dance with rattlesnakes, and rattlers also featured in the beliefs, stories and art of Mayans and Aztecs in Central America. In Egypt the cobra held a particularly strong position and appeared on the head-dress of the Pharaoh while Thais and Sri Lankans associate the cobra closely with Buddha. In Italy a ceremony where the statue of the Virgin Mary is carried through the streets draped in, admittedly nonvenomous, ratsnakes has been a regular event while the death adder has worked its way into the culture and art of Aboriginal

Australians and the Papuan blacksnake has become the totem snake of the magic men of Papua New Guinea.

None of these ceremonial or cultural uses of snakes, and there are many more, affects their survival as a species or group of species. However, snakes have been found useful for other purposes, especially in Asia where huge numbers of snakes are harvested for their leather, to satisfy a primarily Western desire for snakeskin shoes and handbags, and also for their medicinal properties, a more Eastern belief. Snake blood, snake gall bladders and various other parts of a snake are much desired in Chinese medicine and their wholesale harvesting can have a devastating affect on the local snake populations, and ultimately on rice production since the natural predators of rats are being systematically removed from duty. Records demonstrate that between 45,000-109,000 southern Indonesian spitting cobras are collected for their skins annually, just one species, in one country, in one year. Snake restaurants offering snake meat and snake wine are found throughout Asia and sadly a meal of cobra, selected from the cage, taken out back, butchered and prepared, seems to be almost as much a tourist 'must-do' as visiting the temples. Habu wine, strong spirit with the pickled body of a venomous habu, is a Japanese tonic and finally, as a souvenir it is possible to purchase grossly pumped-up stuffed cobras, both 'coiled for the strike' or wrapped around and doing battle with an equally constipated-looking mongoose. I have even seen these items for sale in international airports but purchasers should be aware that taking this sort of gift home in your carry-on baggage could land you with a hefty fine when you reach your destination since some of the species available are restricted in trade by CITES (Convention for the International Trade in Endangered Species of flora and fauna).

Above: *This shop offering stuffed cobras, pythons and monitor lizards to departing tourists is located at Bali International Airport. Several of these species are controlled by CITES regulations.*

Tourists to Asia or North Africa also hope to photograph a snake charmer at work. Some snake charmers care for their charges but others are less concerned for their welfare and sew up the snake's mouths to prevent them from biting, or pull their fangs with pliers or a bitten cloth yanked out of the mouth, though this latter method does not prevent snakebite. In India snake charming is more rarely seen due to strict new laws.

Snakes are also affected by environmental factors such as habitat destruction that may remove their rainforest cover or eliminate their preferred prey species. Some snakes adapt too well to change, the American lanceheads being an example, but others, such as the bushmasters, do not adapt to massive alterations in their surroundings. When Costa Rica started to clear fell forest the terciopelo, a highly venomous lancehead, was able to adapt and thrive in the newly created farmland where it fed on rats attracted by the increase in available food, the result being an annual increase in snakebites from 200 to 400. The same situation occurred in Sri Lanka when an area in the north was irrigated for rice production and the local Russell's vipers sent snakebite statistics through the roof. These are adaptable and ubiquitous species but many others are not able to survive and may disappear.

Roads are a major threat to many organisms but snakes in particular fall foul of their allure. A warm black tarmac road must seem an attractive place for a snake out hunting on a wet night. Lingering too long, or simply trying to cross the road may result in the snake being killed by a passing vehicle. Even fast-moving diurnal snakes are killed on roads. Highly alert and active Montpellier snakes in Europe or taipans in Papua New Guinea, bask in the early morning on warming roads to achieve their optimum hunting temperature, just as they would on a rocky outcrop, and are frequently killed by cars. I have seen drivers swerve and risk an accident to kill a snake on the opposite carriageway. Long established roads, those near towns, and the land for some distance on either side, are often snake-free zones

Above: *Several hundred snakes were discovered when the author joined the Thai Forestry and Wildlife police on a raid on an illegal reptile meat and skin factory near Bangkok. The large specimens are king cobras.*

because the local serpent population has been exterminated.

A final threat comes from the very people who are concerned about deforestation and would never run over a snake on the road. Illegal collection of valuable snakes, whether for sale or personal ownership, can do immense damage to local serpent populations. The very people who 'love' snakes may be endangering them. That is not to say snakes should not be kept in captivity, – quite the opposite – the captive care and breeding of both common and endangered species is an important conservation tool but the illegal collection of endangered, even protected, species cannot be condoned and is not the action of a true reptile enthusiast.

Snakes are by nature secretive animals that provide few clues to their population status, so their numbers may be pushed close to, or into extinction, without us being aware that it is happening. Many common and endangered species are rarely encountered, quite a number being known from only one or two specimens; whether this is a reflection of their secretive nature or whether they are they actually rare and threatened is difficult to determine. Those at most risk tend to be those species with limited distributions ie. island endemics and montane specialists.

Threatened and endangered venomous snakes

The following are examples of venomous snakes threatened by man in various different ways but it is by no means an exclusive list:

- Atlantic coastal bushmaster (*Lachesis muta rhombeata*) threatened by deforestation of the coastal forests of Brazil.

- Bandy-bandys (*Vermicella* spp.) affected by intensive farming techniques in Australia.

- Milos viper (*Macrovipera schweizeri*) threatened by illegal collection and road-kills on four small Greek islands.

- Broadheaded snake (*Hoplocephalus bungaroides*) threatened by collection of Hawkesbury sandstone 'bushrock' around Sydney, New South Wales.

- Timber rattlesnake (*Crotalus horridus*) threatened by active persecution and habitat loss in eastern USA.

- Aruba Island rattlesnake (*Crotalus durissus unicolor*) threatened on Aruba but the subject of an intensive multi-institution captive breeding program to ensure its survival.

- Mangshan pitviper (*Zhaoermia mangshanensis*) threatened by destruction of its tiny montane forest habitat in China.

- Fiji burrowing snake (*Ogmodon vitianus*) threatened by farming and introduced animals.

- Kenyan mountain viper (*Montatheris hindii*) threatened by its miniscule and fragile distribution and predated by birds-of-prey.

What can be done to conserve snakes?

Education is very important. People must realize that snakes are not out to bite them and prefer to be left alone. They should also realize that it is not their duty or right to kill snakes out of hand. Being a snake is not a crime and even venomous snakes serve a valuable purpose by acting as a natural control of disease carrying, rice-stealing rodents.

The need to understand snakes and the ability to live alongside them are important factors for anyone living in snake country. One classic example is that of the Irula of southern India who used to catch snakes to sell for their skins. Now they use their expert skills to catch and milk snakes for venom research, and then release them again.

It is always much harder to promote conservation of an expressionless 'cold-blooded killer' like a venomous snake than it is a cuddly, appealing, furry small mammal but many snakes do need conservation and protection. Although endangered species are legally protected by CITES and local laws, that does not stop them being killed or collected. For some of those species on the brink there are captive breeding programs and habitat protection projects in place but for many others we simply lack the background information.

Venomous Lizards

Many lizards can inflict painful and bloody bites, especially large iguanas and monitor lizards, and Komodo dragon saliva contains bacteria which cause rapidly fatal septacaemia in their prey, but this is not venom. The North American Helodermatidae contains the only two living venomous lizards. They have stocky bodies, bead-like granular scales, rounded heads and chunky tails that serve as fat-stores. Colouration is either black and pink or brown and yellow. The bulbous venom glands of these lizards are located in the lower jaw, where they bulge visibly. Early investigators failed to recognize them as venom glands, expecting glands in the upper jaws like snakes, and concluded the painful bites were caused by infectious saliva. Venom enters the wound by being drawn by capillarity along sharp-edged, grooved teeth during a chewing bite which may last up to 15 minutes. Although exceedingly painful, there are no authenticated human deaths recorded. The positioning of the venom glands proves that the venom mechanism in these lizards evolved independently from snakes. It is designed for defense, rather than prey capture, since the lizard's prey is harmless. Secretive and nocturnal, Gilas and beaded lizards are rarely encountered, spending much of their time underground. They are endangered by collection, habitat destruction and active persecution and are internationally protected.

Gila Monster *Heloderma suspectum*

The Gila monster, pronounced 'Hila', inhabits scrub-desert from Utah to northern Sinaloa, northwest Mexico. Patterned black with pink transverse bars, two subspecies are recognized, the banded and reticulated Gila monsters.

Range: Southwestern USA to northwest Mexico • Max. length: 0.3–0.5m • Venom: Little known, causes extreme pain and swelling; bites very rare no authenticated fatalities • Habitat: Scrub desert • Prey: Eggs, nestling birds and neonate mammals • Reproduction: Oviparous; 2–13 eggs.

Beaded Lizard *Heloderma horridum*

This lizard, named for its bead-like body scales, is known as 'escorpion' in Mexico. Usually patterned with subdued browns and yellows, specimens from southern Mexico may be almost completely brown. Four subspecies are recognized.

Range: Northwest Mexico to Guatemala • Max. length: 0.8–1.0m • Venom: Little known, causes extreme pain and swelling; bites very rare, no authenticated fatalities • Habitat: Dry deciduous woodland • Prey: Eggs, nestling birds and neonate mammals • Reproduction: Oviparous; 4–15 eggs.

"The ranges of the desert-dwelling Gila monster and woodland-dwelling beaded lizard overlap by some 50km in southern Sonora. I set myself the challenge of finding both species close together and set out to search the arroyas and slopes of southern Sonora. We found two Gilas and three beaded lizards. My most memorable capture occurred while driving back towards Alamos in the evening, after finding and filming a beaded lizard on a Mexican ranch. The dirt road ahead rose directly into the setting sun and when we were some 100m from the crown of the hill a beaded lizard walked across the road, silhouetted against the setting sun. Had our cameras been running it would have been an extremely Jurassic moment. Alamos, Sonora, Mexico

Below left: *The Gila monster (*Heloderma suspectum*) from USA, is a desert dweller that uses its venom for defence.*

Below right: *There are still no authenticated fatalities from the bite of the Mexican beaded lizard (*Heloderma horridum*).*

AMERICAS

A desert rattlesnake sounds a warning, the horse shies away... a banded coralsnake disappears in the leaf-litter as the foraging coati-mundi jumps back in alarm – dangerous snakes advertising their lethal capabilities but preferring to warn rather than strike first. The Americas lack true vipers but there are pitvipers aplenty, no cobras or mambas, but their family is represented by red, yellow and black coralsnakes.

Venomous snakes are found from the Canadian prairies to Patagonian Argentina. In between lie the rainforests of Central America, the Amazon, Choco and Brazilian Atlantic coastal forests, the swamps of the Brazilian Pantanal, bayous of Louisiana, the great deserts of southwestern USA and northern Mexico, and the mountains of the Appalachians, Andes, Rockies, Guianan Shield and Mexican Sierra Madres. All provide homes for venomous snakes.

The distribution of venomous snakes in the Americas is a story of land-bridges, invasions and re-invasions. The Bering land-bridge, linking North America with Asia, permitted the ancestral pitvipers and coralsnakes to reach the new continent during the Miocene era. However, the northern continent was not physically joined to the southern continent until the Pliocene era (three to five million years ago). Prior to this time the only links between the two 'Americas' would have been a chain of volcanic islands so ancestral snakes reached the southern continent by island-hopping or oceanic dispersal. When links were established with the rising of the Panamanian isthmus in the Pliocene it made north-south and south-north radiation of species much easier.

Desert and savanna loving rattlesnakes never really established themselves as rainforest snakes and failed to conquer the Amazon while bushmasters do not like open country and disturbance, precisely the kind of opportunity relished by the ubiquitous lanceheads. Most coralsnakes prefer woodland or forest but a few moved into open desert habitats. Of all the American mainland countries only Chile lacks any front-fanged venomous snakes.

Right: *A coiled western diamondback rattlesnake (*Crotalus atrox*), is one of the abiding images of the Wild West, yet rattlesnakes are found in all but four US States and all but three mainland American countries.*

American Elapids

The coralsnakes are the only American, non-marine, members of the Elapidae, the family so strongly represented by cobras and their kin in Africa and Asia and taipans and tigersnakes in Australasia.

True Coralsnakes

There is a misconception that 'coralsnake' is another name for 'seasnake', possibly because many seasnakes inhabit coral reefs. Coralsnakes are terrestrial snakes and earn their name from the brilliant coral-red bands that characterize many, but not all, of the seventy-plus species. Red is a common colour in American snakes, especially in the highly venomous coralsnakes and their mildly venomous or nonvenomous 'false coralsnake' mimics, but it is rare in snakes from other continents. Coralsnakes are found in every mainland country except Canada and Chile, and also on a few continental islands. Most species are found in rainforests but 'corals' also inhabit swamps, rivers, woodlands, grasslands and even deserts. They are primarily secretive nocturnal snakes with small eyes and they are rarely encountered unless uncovered when rolling a log. Although highly venomous, coralsnakes feature little in snakebite statistics, accounting for only 0.9% of serious bites in Latin America. Most coralsnakes belong to the large genus *Micrurus* but the Sonoran coralsnake and the four species of slender coralsnakes, are placed in two different genera. The numerous coralsnakes of genus *Micrurus* may be divided into subgroups depending on the arrangement of their red and black bands. White or yellow rings may be absent in some species or populations. Monadal coralsnakes have a single black ring between each pair of red rings (red/yellow/black/yellow/red). Triad coralsnakes have three black rings in a sequence between each pair of red rings (red/black/yellow/black/yellow/black/red).

Sonoran coralsnake *Micruroides euryxanthus*

The only coralsnake occurring in southwestern USA, this snake is the sole member of its genus. It inhabits a wide variety of different habitats but prefers dry river beds, where it preys mostly on reptiles. A curious defense of this species involves the expulsion of air through the cloaca (anal opening) known as cloacal popping. Three subspecies are recognized.

Range: Southwestern USA and northwestern Mexico • Max. length: 0.3–0.55m • Venom: Postsynaptic neurotoxin and myotoxin; snakebites but no fatalities • Habitat: Desert to dry deciduous woodland • Prey: Snakes and lizards • Reproduction: Oviparous; 2–3 eggs.

Below: *The Sonoran coralsnake (*Micruroides euryxanthus*) occurs in southwest USA/northwest Mexico. It is probably the least dangerous North American coralsnake.*

Guianan slender coralsnake *Leptomicrurus collaris*

Although they are coralsnakes, these snakes do not exhibit the brightly coloured bands of their larger relatives, being almost black above apart from a yellow nape band. The undersurfaces are patterned with regular yellow spots, which may be visible on the lower flanks. The underside of the tail is red-orange. Defensive display involves rolling over and exposing the underside of the tail. Two subspecies are recognized.

Range: Eastern Venezuela, the Guianas and northern Brazil • Max. length: 0.4–0.45m • Venom: Postsynaptic neurotoxin and myotoxin; no snakebite reports • Habitat: Lowland and low montane rainforest • Prey: Small slender lizards and blindsnakes • Reproduction: Oviparous; probably 1 egg • Similar species: Andean slender coralsnake (*L. narducci*)

Monadal Coralsnakes

Monadal coralsnakes are those that are patterned with a single black band between each pair of red bands, i.e. red/yellow/black/yellow/red.

Eastern coralsnake *Micrurus fulvius*

The only coralsnake in eastern USA, the eastern coralsnake is a widespread, versatile but secretive species that is far less frequently encountered than the pitvipers that inhabit the same southeastern states. The coralsnake's nocturnal to crepuscular (active at twilight) habits may help to keep snake and humans apart. They do not bask in the sun like pitvipers. Their patterning consists of similar-sized black and red spaces separated by narrow yellow bands, with a broad yellow band across the rear of the black head and neck. Although not as commonly implicated in snakebite accidents as rattlesnakes and their kin, it was reportedly an eastern coralsnake that caused the first fatality of the American Civil War.

Range: Southeastern USA • Max. length: 0.7–1.0m • Venom: Postsynaptic neurotoxin and myotoxin; snakebites but fatalities rare • Habitat: Many habitats, especially tropical hammocks in Florida • Prey: Small snakes and slender lizards • Reproduction: Oviparous; 2–12 eggs • Similar species: Texas coralsnake (*M. tener*) and milksnake (*Lampropeltis triangulum*).

Painted coralsnake *Micrurus corallinus*

The usual painted coralsnake pattern consists of a monad of a blank ring between narrow yellow/white rings, separated from the next monad by a broad red interspace but some specimens are unringed, being all red apart from black and yellow rings on the head and tail. Preying on a wide variety of slender vertebrates, painted coralsnakes are not adverse to eating their own species.

Range: Atlantic coastal of Brazil to Uruguay and northern Argentina • Max. length: 0.65–0.95m • Venom: Postsynaptic neurotoxin and myotoxin; snakebites recorded, including fatalities • Habitat: Tropical and subtropical deciduous, and coastal forest • Prey: Amphisbaenians and caecilians, also snakes including cannibalism • Reproduction: Oviparous; 2–10 eggs • Similar species: Peruvian coralsnake (*M. peruvianus*).

Top: *Although a small species, the Eastern coralsnake (*Micrurus fulvius*) has caused human fatalities.*

Above: *The Painted coralsnake (*Micrurus corallinus*) is a typical monadal coral with patterning arranged in red/yellow /black/yellow/red bands. This specimen is in pre-slough.*

Triad Coralsnakes

These snakes are patterned with three black rings between each pair of red rings, i.e. red/black/yellow/black/yellow/black/red.

South American coralsnake *Micrurus lemniscatus*

Although not recorded from the central Amazon and the truly arid parts of Brazil, the four subspecies are distributed as far afield as the Andean foothills and the Atlantic coastal forests. This is the typical South American coralsnake – a moderate sized, triad-patterned, secretive forest dweller that preys on slender terrestrial vertebrates.

Range: Widespread throughout South America. • Max. length: 0.6–1.1m • Venom: Postsynaptic neurotoxin and myotoxin; snakebites recorded • Habitat: Lowland and low montane rainforest, savanna and cleared areas • Prey: Small snakes, slender lizards, amphisbaenians, caecilians, eels and slender fish • Reproduction: Oviparous; 4–10 eggs • Similar species: Thread coralsnake (*M. filiformis*) and false coralsnake (*Erythrolamprus aesculapii*).

"When I worked on the Royal Geographical Society Maracá Rainforest Project in 1987-88 I found the S.American coralsnake the most frequently encountered coralsnake inhabiting both dry forest and savanna habitats, but I found its false coralsnake mimic much more common.
Ilha Maracá, Roraima, Brazil.

Amazonian coralsnake *Micrurus spixii*

This coralsnake holds the record for the largest coralsnake recorded at 1.6m but most specimens fall far short of this length. Four subspecies are recognized which may be termed lower Amazon, central Amazon, upper Amazon and Bolivian. The triad banding of the Amazonian coralsnake comprises very few triads because the yellow or white rings, which are narrow in most other species, are broader than the black rings and in some specimens broader than the red bands.

Range: Amazon and Orinoco basins and eastern Andean foothills • Max. length: 0.8–1.6m • Venom: Postsynaptic neurotoxin and myotoxin; snakebites recorded • Habitat: Rainforest and secondary forest • Prey: Small to medium-sized lizards and snakes, including lanceheads • Reproduction: Oviparous; 6–12 eggs • Similar species: Venezuelan coralsnake (*M. isozonus*) and false coralsnake (*Erythrolamprus aesculapii*).

Aquatic coralsnake *Micrurus surinamensis*

The aquatic coralsnake is another large species, with a distribution that overlaps and exceeds that of the Amazonian coralsnake. It can also be distinguished from other coralsnakes by its red head with every scale edged with black. The most aquatic of coralsnakes, this species has a broad, flattened head which may be a reflection of its freshwater lifestyle. Its prey consists mostly of fish species and mimics tend to be aquatic colubrid snakes. Two subspecies are recognized.

Range: Guianas and Amazon and Orinoco basins to eastern Andean foothills • Max. length: 0.6–1.35m • Venom: Postsynaptic neurotoxin and myotoxin; snakebites recorded, including fatalities • Habitat: Rainforest and wet forest along rivers and swamps • Prey: Swamp eels, knifefish and catfish • Reproduction: Oviparous; 5–13 eggs • Similar species: Harmless watersnakes (*Helicops* and *Hydrops* spp.).

Top: *The South American coralsnake (*Micrurus lemniscatus*) is a typical triad coralsnake with patterning arranged in red/black/yellow/black/yellow/black/red bands.*

Left: *The Amazonian coralsnake (*Micrurus spixii*), named for the German zoologist and Amazonian explorer Johann Baptist von Spix (1781-1826), holds the record for the longest coralsnake at 1.6m.*

False Coralsnakes

Throughout the entire range of coralsnakes there are nonvenomous and/or mildly venomous snakes that closely resemble them and have long been termed 'mimics' or 'false coralsnakes'. North American coralsnakes can be distinguished from the harmless milksnakes by the rhyme 'red to yellow, kill a fellow; red to black, venom lack' (or 'friend to Jack', which is a reference to the milksnake's venomous snake-eating habits). By examining the snake and determining whether the red band is in contact with the black or yellow band it is possible to distinguish a dangerous coralsnake from a harmless milksnake. The rhyme does not work in Latin America where coralsnake patterning is more diverse, and there are also many more species of mimics ranging from the simple red-and-black-banded South American pipesnake (*Anilius scytale*) to elaborately patterned false coralsnakes. In some parts of Latin America the coralsnake and the false coralsnake even demonstrate startlingly similar local variations in their patterning, as if one were copying the other, but which is the 'mimic' and which is the 'model'?

Milksnake *Lampropeltis triangulum*

This is the mostly widely distributed American snake. All but two of its 25 species resemble mondal coralsnakes except that their band order is 'red to black, venom lack'. Much of its N. American range lies outside the range of any coralsnake, thereby throwing doubt on the mimicry argument. The bright patterning might simply confuse the predator and provide a chance for the snake to escape.

Range: Southeastern Canada and USA to Andean Colombia Ecuador • Max. length: 0.38–1.9m • Venom: Nonvenomous. • Habitat: Temperate grassland to tropical rainforest and desert • Prey: Snakes, lizards, small mammals • Reproduction: Oviparous; 5–16 eggs.

Right: *South American False coralsnakes, such as* Erythrolamprus aesculapii, *are mildly venomous, rear-fanged snakes with larger eyes than coralsnakes. The small, square loreal scale between the eye and nostril is absent from all corals.*

False coralsnake *Erythrolamprus aesculapii*

This is one of the commonest false coralsnakes. It even adopts the coralsnake defensive behaviour of body flattening and tail rolling. It exhibits the safe 'red to black, venom lack' patterning but unfortunately so do many dangerous South American coralsnakes. The proportionally larger eye and the presence of a loreal scale will identify this rear-fanged species.

Range: Amazon Basin • Max. length: 0.6–0.8m • Venom: Mildly venomous but little known • Habitat: Rainforest and secondary forest • Prey: Small snakes • Reproduction: Oviparous: 5 eggs recorded.

Below *The North American nonvenomous Milksnake (*Lampropeltis triangulum*) can be distinguished from the coralsnakes by the rhyme 'red to black, venom lack'. This is the Pueblan subspecies (*L. t. campbelli*)*

American Pitvipers

All American vipers are pitvipers. They possess heat-sensitive facial pits which aid them in the capture of warm-blooded prey. There are no 'true vipers 'in the Americas.

Copperheads, Cottonmouths and Cantils

These American pitvipers are most closely related to ancestral pitvipers which crossed the Bering land-bridge from Asia 24 million years ago. All four species have bright yellow tail tips when they are young, which they use as caudal lures to attract prey to within strike range. As they age and the tip colour becomes more subdued, they use the tail defensively, thrashing it noisily in leaf litter to deter interference.

Copperhead *Agkistrodon contortrix*

The Copperhead is a common snake throughout eastern USA, but, oddly, not peninsular Florida. An attractive snake, with patterning comprising of a mixture of russet brown saddles and subtle autumnal shades, it is the least dangerous of the group with few deaths documented, although children may be more at risk. There is considered sufficient morphological variation across the range to recognize five subspecies and venom toxicity may also vary within the species.

Range: Eastern USA, New England to Texas and northeast Mexico • Max. length: 1.0–1.3m • Venom: Weak coagulants, local pain and swelling usual; many snakebites, but deaths very rare • Habitat: Dry deciduous pine-oak woodlands and rocky hillsides • Prey: Mammals and other terrestrial vertebrates. • Reproduction: Viviparous; 2–18 neonates.

"I received a graze down the thumb from a juvenile specimen but it did not appear to have broken the skin so I disregarded it. Some hours later I was in my car and my thumb appeared to have 'gone to sleep'. It being winter I blamed the cold and turned on the car heater, but could not get the feeling back into my thumb. Only then did I realize I had been mildly envenomed four hours earlier and that was why my thumb, from tip to ball, appeared dead down one side. The symptoms wore off in a couple more hours.

Cottonmouth *Agkistrodon piscivorus*

The name 'cottonmouth' refers to the white interior of the mouth that is revealed when the pitviper gaps in a threat display. The alternative name of 'water moccasin' is a Native American name, a reference to the aquatic habits of this heavy-set pitviper. Juveniles have olive-brown saddles on a darker background but with increasing age the pattern becomes obscured by dark pigment until the cottonmouth is virtually black. Often mistaken for a harmless American watersnake, the three subspecies of cottonmouths cause numerous snakebites and some deaths.

Range: Southeastern and south-central USA • Max. length: 1.0–1.5m • Venom: Procoagulants, myotoxins, pain and swelling may be severe, many snakebites, some deaths • Habitat: Swamps, bayous and rivers • Prey: Fish, frogs and other vertebrates • Reproduction: Viviparous; 4–11 neonates • Similar species: Harmless watersnakes (*Nerodia* spp.).

Left: *The woodland-dwelling Copperhead (*Agkistrodon contortrix*) is the least venomous species in the genus with only two recorded fatalities in recent times.*

Mexican cantil *Agkistrodon bilineatus*

Cantil patterning consists of brown, pink or orange saddles, edged with white, over a darker brown background, and a series of five white stripes radiating backwards from the snout, down the chin, each side of the upper jaw, and through the eye. The name 'cantil' is believed to mean 'viper-snake' in Mayan. Three subspecies are recognized for Mexico–Guatemala, Yucatán and Honduras–Costa Rica. The most venomous of the group, cantil bites have caused the death of humans and horses from 20 minutes to a few hours later.

Range: Pacific Coast of Mexico to Costa Rica and Yucatán peninsula • Max. length: 0.8–1.38m • Venom: Procoagulants, myotoxins and haemorrhagins; few snakebites but rapid deaths reported • Habitat: Tropical deciduous woodland • Prey: Mammals and other terrestrial vertebrates • Reproduction: Viviparous; 3–20 neonates • Similar species: The isolated and grey-yellow liveried Taylor's cantil (*A. taylori*), from northeastern Mexico, was originally a subspecies but now has species status. Its patterning has led to the alternative name of Ornate cantil.

"I have to confess to a couple of bites from this species, but being juvenile specimens they did not inject a great deal of venom and the effects were limited to localized pain and swelling, as well as pain in the axillary lymph node, which wore off within a few hours.

Top: *The Cantil (*Agkistrodon bilineatus*) is the most venomous member of the genus. Characterised by the five white head stripes, it has caused rapid human fatalities.*

Above: *The Cottonmouth or Water Moccasin (*Agkistrodon piscivorous*) is the most aquatic of all American pitvipers. It is named for its brilliant white mouth interior.*

Lanceheads

There are about 30 species of lanceheads. These primarily terrestrial, but in some cases semi-arboreal, pitvipers are found from Mexico to Argentina and on a few islands. There are four island endemics. Several lanceheads are known by their tongue-twister local names, terciopelo, jararaca, jararacussu, cotiara and urutu, rather than by the lancehead suffix. The lanceheads are the ubiquitous venomous snakes, highly adaptable species which moves into agricultural land or newly cleared habitats and produce large numbers of venomous offspring. The lanceheads, as a group, are responsible for about 90 per cent of all serious snakebites in Latin America, and since their venom is highly cytotoxic even survivors may be subjected to massive tissue damage leading to debilitating injuries or amputation.

Common Lancehead *Bothrops atrox* and Terciopelo *Bothrops asper*

The common lancehead kills and maims more people than any other species of snake in the nine South American countries where it occurs, notably in the Amazon region. The terciopelo is also an extremely dangerous snake within its range. It was formerly treated as a subspecies of the common lancehead but is now recognised as a valid species, distinct from its southern-relative. Adapting to disturbance well, they are often encountered around buildings and frequently come into contact with humans. Although adults are primarily terrestrial, juveniles are adept at climbing and may be encountered at hand or face height. Patterning may be highly variable causing confusion in regions with more than one lancehead species. I have caught the common lancehead in Brazil, Guyana and Venezuela, but have also come to respect a large terciopelo as a very aggressive, agile and dangerous animal.

Range: Mexico to South America on the Pacific coast to Ecuador, on the Caribbean, to Venezuela and Trinidad (*B. asper*); Amazonian countries (*B. atrox*) • **Max. length:** 0.75–1.25m (*B. atrox*); 1.8–2.5m (*B. asper*) • **Venom:** Procoagulants, myotoxins, cytotoxins, many snakebites, debilitating injuries and fatalities • **Habitat:** Most habitats • **Prey:** Small mammals, birds, lizards and frogs • **Reproduction:** Viviparous; 8–33 neonates (*B. atrox*); 25–70 neonates (*B. asper*).

"The terciopelo is unpredictable and very dangerous. When moving at speed it changes direction by lifting the entire front third to one half of its body off the ground before dropping it in the new direction so that when it changes direction it has already traveled up to 1.0m in that direction. This makes pursuing and capturing them on forest hillsides and in tangled jungle vegetation rather risky. The head of the terciopelo is also the most sinister snake's head I have seen, it even looks angry when it is asleep, unnoticed in leaf litter. Honduras, Costa Rica and Trinidad.

"Biologists found a juvenile terciopelo coiled on the concrete outside an ecological station building. Although they laid out a loop of coloured cord around the little pitviper, to indicate its position, it was incredible how its cryptic patterning made it virtually invisible on plain grey concrete, forcing one to concentrate hard in order to see it. If the snake blended in on concrete it would be invisible in leaf litter. **La Selva Reserve, Costa Rica.**

Jararaca *Bothrops jararaca*

The jararaca is one of several lanceheads responsible for the high snakebite statistics in the highly populous region around São Paulo and Rio de Janeiro and on the island of São Sebastião. The dwarf population from Alcatraz island, off São Sebastião, has recently been described as a new species, *B. alcatraz*.

Range: Southeastern Brazil • Max. length: 2.0–2.4m • Venom: Procoagulants, myotoxins, cytotoxins; frequent cause of serious snakebites, often fatal • Habitat: Deciduous woodland and open cultivated country • Prey: Small mammals, birds, lizards and frogs • Reproduction: Viviparous; 18–22 neonates • Similar species: Jararacussu (*B. jararacussu*) from Brazil.

Patagonian lancehead *Bothrops ammodytoides*

The Patagonian lancehead, with its upturned snout, short, stout body and brown or grey blotched patterning, bears a striking resemblance to the nose-horned viper of southeastern Europe. The most interesting aspect of this little pitviper is its claim to be the southern-most snake in the world. Occurring in southern Patagonia, its range extends further south than any other South American, South African or Tasmanian snake. Being viviparous is a definite advantage because eggs laid in this, often harsh, environment would perish quickly.

Range: Patagonia, Argentina • Max. length: 0.7–1.0m • Venom: At least procoagulants, probably highly dangerous but no records of snakebites • Habitat: Temperate pampas grasslands and coastal sand dunes • Prey: Lizards, sometimes frogs • Reproduction: Viviparous; litter size unknown • Similar species: São Paulo lancehead (*B. itapetiningae*) from Brazil and European nose-horned viper (*Vipera ammodytes*).

Opposite: *The Terciopelo (*Bothrops asper*) is one of the most aggressive and dangerous snakes in the Americas.*

Above: *The Jararaca (*Bothrops jararaca*) is one of numerous confusingly similar Brazilian lanceheads responsible for many serious snakebites.*

“I had been looking for this little lancehead in perfect habitat all over Patagonia, without success, so decided to reconsider my methodology. I thought of the British adder. There are many perfect adder habitats in the UK without adders, but where adders do occur you can usually find several. I reasoned that was because of availability of hibernacula, potential mates and prey, so I looked for sheltered habitats with abundant lizards. I found a location with plenty of lizards on a cliff top and started to look closely at sheltered gaps in the vegetation. Within 20 minutes I found a gravid (pregnant) female lancehead and within another ten my Brazilian colleague had found an adult male. **Peninsula Valdes, Patagonia, Argentina.**

Golden lancehead *Bothrops insularis*

The golden lancehead exists on an isolated Brazilian island and may have had no contact with mainland pitvipers for between 100,000 to one million years. In the absence of small mammals it has evolved to feed on birds, which it captures when they land to feed on fruit at the base of trees. The lancehead has long fangs to penetrate the plumage and the most toxic venom of any American snake. Unlike mainland lanceheads, which strike, inject venom into and then release their mammalian prey, only to track it down later when it is safely dead, the golden lancehead must retain hold of the bird and kill it quickly. If it lets go and the birds flutters away, the snake will not find it again. Consequently, its

Opposite: *The Patagonian lancehead (*Bothrops ammodytoides*) is the southern-most snake in the world, being distributed deep into Argentine Patagonia.*

Above: *The Golden lancehead (*Bothrops insularis*) of Ilha Queimada Grande is not only the most isolated Latin American venomous snake, but also the most venomous.*

venom is three to five times more toxic than that of any mainland venomous snake. The Ilha Queimada Grande (IQG) lanceheads also exist as three sexes, males, females (now rare) and inter-sex females (females with inactive male genitalia). There are estimated to be some 5000 golden lanceheads in the forest of IQG which has an area of around 430,000 sq.m.
Range: Ilha Queimada Grande, off São Paulo, Brazil.

• Max. length: 0.7–1.2m • Venom: Procoagulants, myotoxins, cytotoxins; no snakebites on record but highly venomous • Habitat: Island forest • Prey: Birds • Reproduction: Viviparous; 2–10 neonates • Similar species: None, though possibly closest to jararaca on mainland and newly described Alcatraz Island lancehead (*B. alcatraz*).

"We visited IQG with herpetologists from the world famous Instituto Butantan and the University of São Paulo who were conducting a study of the golden lanceheads. I was surprised how many lanceheads could be found in a short time, I found 16 in little more than one hour, on the ground and in the trees.
Ihla Queimada Grande, Brazil.

Forest-pitvipers

Some experts consider the six species of South American forest-pitvipers, of genus *Bothriopsis*, part of the lancehead genus *Bothrops*, which contains terrestrial, arboreal and semi-arboreal species. The forest-pitvipers have been treated separately here as a comparison with the Central American palm-pitvipers of genus *Bothriechis*.

Two-striped forest-pitviper *Bothriopsis bilineata*

This small arboreal pitviper is green or cyan, with black speckling, faint dark blotches on the back and yellow lips. This is the second most dangerous snake in the Amazon after the common lancehead (*Bothrops atrox*). I examined a small specimen of the eastern subspecies, which had killed an man in Alagoas, Brazil and a friend was seriously bitten by a specimen of the western subspecies in Ecuador. Even small venomous snakes can be lethal.

Range: Western Amazon from Colombia to Bolivia, Venezuela to the Atlantic forests of Brazil • Max. length: 0.7–1.0m • Venom: Procoagulants, anticoagulants and haemorrhagins; serious snakebites and fatalities known • Habitat: Lowland primary and secondary rainforest • Prey: Tree frogs, lizards, small birds and mammals • Reproduction: Viviparous; 4–15 neonates • Similar species: Werner's false lancehead (*Xenodon werneri*).

Speckled forest-pitviper *Bothriopsis taeniata*

The speckled forest-pitviper is patterned with a complicated design of speckled green and brown bands in a lichen-like pattern. The underside is marked by a row of white spots at the end of the enlarged ventral scales. Its preferred habitat is forest-edge where its cryptic patterning enables it to secret itself away in vines or other vegetation to ambush prey.

Range: Colombia, around Amazon Basin, through Peru and Brazil, to Guianas and Venezuela • Max. length: 1.0–1.5m • Venom: Anticoagulants and haemorrhagins; serious snake-bites recorded, but no records of fatalities • Habitat: Lowland primary and secondary rainforest • Prey: Treefrogs, lizards, small mammals • Reproduction: Viviparous; 7–17 neonates.

Left: *The Speckled forest-pitviper (*Bothriopsis taeniata*) is a cryptically patterned species that blends into vine entanglements in forest-edge habitats.*

Above: *The Two-striped forest-pitviper (*Bothriopsis bilineata*) may be small and appear less threatening that a larger terrestrial lancehead but it has a proven history of causing human fatalities.*

Palm-pitvipers

The nine Central American palm-pitvipers of genus *Bothriechis* are separate from the lancehead genus *Bothrops*. Most have limited distributions but several are more widespread.

Eyelash palm-pitviper *Bothriechis schlegelii*

This snake exists in a startling array of colour morphs, from cryptic lichen-patterned specimens, through plain brown or green to bright yellow-gold forms, sometimes termed 'oropel'. This is the most widespread of the Central American palm-pitvipers and the only species to enter South America. It can easily be distinguished from most other Central American pitvipers by a row of raised scales above the eye – the 'eyelashes'.
Range: Yucatán peninsula to Ecuador and western Venezuela • Max. length: 0.6–0.9m • Venom: Procoagulants and haemorrhagins; mixed reports of snakebites, rare fatalities have been reported • Habitat: Lowland primary and secondary rainforest • Prey: Lizards, treefrogs, small birds and rodents • Reproduction: Viviparous; 2–20 neonates • Similar species: Blotched palm-pitviper (*B.supraciliaris*), a former subspecies.

"Exploring the Costa Rican rainforest at night I encountered numerous eyelash palm-pitvipers in ambush positions in the low branches. Colours varied from green to bright yellow-gold and even a white specimen. Costa Rica.

Right: *The Black-speckled palm-pitviper (*Bothriechis nigroviridis*) is one of several confusing C. American green palm-pitvipers.*

Above: *The golden phase of the Eyelash palm-pitviper (*Bothriechis schlegelii*), named for German naturalist Hermann Schlegel (1804-1884), is known as Oropel in Costa Rica.*

Black-speckled palm-pitviper *Bothriechis nigroviridis*

This slender, arboreal, pitviper is green above, every scale edged with black and black mottling on head. It could easily be confused with other green palm-pitvipers were it not for the distinctive black markings. An inhabitant of undisturbed montane forests, it is a rare species which does not adjust well to deforestation.
Range: Costa Rica and western Panama • Max. length: 0.6–0.8m • Venom: Procoagulants and haemorrhagins; mixed reports of snakebites and rare fatalities have been reported • Habitat: Montane and cloud forest • Prey: Lizards, treefrogs, small birds and rodents • Reproduction: Viviparous; 4–8 neonates • Similar species: At least six species of green *Bothriechis*.

Jumping and Hognosed Pitvipers

Twenty-five, primarily terrestrial, pitvipers, formerly placed in the lancehead genus *Bothrops*, have been removed to five genera resurrected or newly created to accommodate them.

Central American jumping pitviper *Atropoides mexicanus*

The stout-bodied, jumping pitvipers of genus *Atropoides* comprise five species distributed from Mexico to Panama. Their defensive posture consists of forming a tight coil with the head in the centre, mouth agape. These small snakes strike with so much force that they often flip their entire bodies forwards, hence their common name. The Central American jumping pitviper is the most wide-ranging of the five species and its venom varies between the Caribbean and Pacific populations.

Range: Central America, Southern Mexico to Panama • Max. length: 0.5–0.9m • Venom: Pro- and anticoagulants, possibly haemorrhagins, poorly known • Habitat: Tropical and cloud forest, and wooded savanna • Prey: Rodents, large grasshoppers and lizards • Reproduction: Viviparous; 13–36 neonates • Similar species: Mexican jumping pitviper (*A. nummifer*) of which *A. mexicanus* was once a subspecies.

Amazonian toad-headed pitviper *Bothrocophias hyoprora*

The stout-bodied, terrestrial, toad-headed pitvipers of genus *Bothrocophias* comprise five species distributed from Colombia to Bolivia. Although the Amazonian toad-headed pitviper occurs within the range of similar terrestrial pitvipers, its sharply upturned snout serves to distinguish it from other species. Toad-headed pitvipers bear their strongest resemblance to the hognosed pitvipers of Central America.

Range: Western Amazon, Colombia to Bolivia • Max. length: 0.5–0.8m • Venom: Procoagulants and haemorrhagins; snakebites with child fatalities have been recorded • Habitat: Rainforest • Prey: Lizards and rodents. • Reproduction: Viviparous; 4 neonates • Similar species: Small-eyed toad-headed pitviper (*B. microphthalmus*) of the Andes.

Godman's montane pitviper *Cerrophidion godmani*

The moderately slender, terrestrial, montane pitvipers of genus *Cerrophidion* comprise one widespread and fairly well-known, species and three species on isolated Mexican mountain ranges. Godman's montane pitviper inhabits a variety of montane habitats through its wide range, sometimes in areas of high rainfall and sometimes in particularly arid locations.

Opposite: *The stout-bodied Jumping pitviper (*Atropoides mexicanus*) throws itself forwards when it strikes.*

Above: *Godman's pitviper (*Cerrophidion godmani*) is the most widespread and best known of the four Central American montane pitvipers.*

Range: Central America, southern Mexico to Panama. • Max. length: 0.55–0.75m • Venom: Procoagulants, haemorrhagins, myotoxins, possible cytotoxins; serious bites but no fatalities. • Habitat: Wet and dry montane forest, and cloud forest • Prey: Rodents, grasshoppers and some lizards • Reproduction: Viviparous; 2–12 neonates • Similar species: Tzotzil montane pitviper (*C. tzotzilorum*) from Chiapas, Mexico.

Rainforest hognosed pitviper *Porthidium nasutum*

The stout-bodied, terrestrial, hognosed pitvipers of genus *Porthidium* comprise nine species from Central America to northern and northwestern South America. Some species are widespread, others exhibit very limited ranges. Hognosed pitvipers can be distinguished from other terrestrial pitvipers within their range by the raised snout, especially in the case of the rainforest hognosed pitviper which possesses a sharply upturned snout appendage and also raised ridges over the eye.

Range: Southern Mexico to Ecuador • Max. length: 0.4–0.6m • Venom: Procoagulants and haemorrhagins; many snakebites causing local pain, few fatalities • Habitat: Lowland rainforest and dry forest • Prey: Lizards, frogs and birds. • Reproduction: Viviparous; 2–15 neonates • Similar species: Slender hognosed pitviper (*P. ophryomegas*) from Guatemala to Costa Rica.

"On an expedition into the remote Mosquito Coast in 1985. I discovered the best time for finding snakes was from 06:30-08:30 in the morning (PST - peak snake time). This was generally the time we had broken camp and were on the move through the jungle and this was when I found two rainforest hognosed pitvipers and an eyelash palm-pitviper (*Bothriechis schlegelii*) all too chilled to be alert to our approach. After this time the snakes were warming up and easily able to elude someone trying to capture them whilst carrying a heavy rucsac. Mosquito Coast, Honduras.

Mexican horned pitviper *Ophryacus undulatus*

The two moderately stout, semi-arboreal/terrestrial horned pitvipers of genus *Ophryacus* are fairly poorly known but they can be easily distinguished from all other Central American pitvipers, apart from the eyelash palm-pitviper, by the raised, horn-like scales over their eyes. The Mexican horned pitviper is fairly variable in coloration, ranging from pale grey to lichen green, with a slightly darker zigzag pattern down the back. This is a species found at moderate elevations in the southern mountain ranges of Mexico.

Range: Southern Mexico. • Max. length: 0.55–0.7m • Venom: Procoagulants and haemorrhagins; no snakebites recorded • Habitat: Pine-oak and cloud forest • Prey: Lizards and rodents • Reproduction: Viviparous; 6–13 neonates • Similar species: Black-tailed horned pitviper (*O. melanurus*) and eyelash palm-pitviper (*Bothriechis schlegelii*) also from southern Mexico.

Bushmasters

Among the venomous snakes of the Americas the bushmasters are unique. These huge, primarily nocturnal forest pitvipers are the longest vipers in the world and the only American pitvipers to lay eggs. The human inhabitants of the rainforests also hold them in a combination of awe and fear and there are a number of stories that emphasize the respect with which they view this deadly viper. In Brazil if workers hear the 'whistle of the surucucu' they will stop working immediately and go back to camp. Not to do so is said to invite death. Our forestry workers frequently heard bushmasters whistle when asked to cut a trail through particularly dense bush on Ihla Maracá, Roraima, northern Brazil, but in seven months there I never found a bushmaster. Bushmasters are also said to hunt in pairs, and to be attracted to campfires; neither of which are likely to be true.

True rainforest snakes, requiring annual precipitation rates of 2000-6000mm, bushmasters are rarely found in dry forest. They dislike disturbance. When rainforest is cleared and replaced with agricultural land or overgrown with secondary growth, the bushmaster will soon disappear and be replaced by more ubiquitous, and potentially more dangerous, lanceheads (*Bothrops* spp.). Bushmasters are much feared and although hospital admission statistics show that it is responsible for only 0.01 per cent of recorded Latin American snakebites, a bite received in the jungle will almost certainly be fatal and be omitted from the official statistics because bushmaster victims may never reach hospital. Bushmasters may be threatened with extinction due to the felling of their native habitat. This is especially the case with the Atlantic coastal bushmaster. All bushmasters were considered one species, with four subspecies, but recent studies recognize three valid species, and a single subspecies. A fourth species, the Chocoan bushmaster (*Lachesis acrochorda*) has now been recognized from Panama, Colombia and Ecuador.

Bushmasters are attractively patterned with a series of dark diamond-shaped markings over the centre of the back, overlying a yellowish or pinkish ground colour. Each side of the dark marking is also centred with light pigment. The body is stout with a raised vertebral ridge and the head is rounded, rather than angular like other pitvipers. The spinous texture of the scales of the back resulted in the local name of *pico de jaca* or 'spines of the jack fruit.' A curious spine-like structure on the end of the tail, resulting from the fragmentation of the last few scale rows, earned the bushmaster the alternative name of 'mute rattler' and contributed to the scientific name of *Lachesis muta*. The aggression of the bushmaster has been overstated; the black-headed bushmaster has been called aggressive and may be the reason the species is locally known as *matabuey* or 'ox-killer', but Bushmasters do not pursue people, although an angry adult can put on an impressive display, raising its body, inflating its neck and vibrating its tail on the leaf litter. Bushmasters are largely sedentary snakes that remain in mammal burrows for long periods, lie motionless in ambush for days and may live for more than 30 years.

"Driving through Venezuela I saw what looked like a dead boa constrictor beside the road. When I stopped to examine the body I discovered it was a bushmaster and it had not been killed by a car, it had been shot. Having insufficient preservative to fix its as a museum specimen I decided to skin it at the next break in our journey. When we stopped I found somewhere quite and set to work but before long an audience had gathered. I removed the skin and the carcase lay on the sand, attracting the attention of the ants. A

young Amerindian girl asked for the ant-covered carcase for her grandmother who would use it to make potions. The use of animal body parts in tribal medicine and tonics is a common occurrence throughout the world and the feared bushmaster may be thought to hold hidden powers. El Dorado, Bolivar, Venezuela.

"Some 95 per cent of the Atlantic coastal forest in Brazil has been destroyed, with the result that all indigenous wildlife has become endangered. One of my early films involved a search for the elusive Atlantic coastal bushmaster (*L. muta rhombeata*), a subspecies of the South American bushmaster, in the company of two Brazilian experts who had only found eight live specimens themselves in their combined careers - this is a rare snake. We searched intensively for ten dry days and nights in what remained of prime habitat in three Atlantic states. On the final afternoon of filming it rained for a few hours so we decided on a final search that evening. I was about to call the search off at around midnight when a member of the crew called me, urgently, and we found a sub-adult male bushmaster lying beside a fallen tree in the jungle. It was not so much eleventh hour as twenty-fourth hour. Quebrangulo, Bahia, Brazil.

Opposite: *The South American bushmaster (*Lachesis muta*) is found from Trinidad to the Atlantic coastal forests of Brazil.*

Central American bushmaster *Lachesis stenophrys*
Range: Caribbean coast from Nicaragua to Panama • Max. length: 2.0–3.3m • Venom: Pro- and anticoagulants, and haemorrhagins; snakebites rare but fatality rate high and rapid • Habitat: Tropical rainforest and riverine gallery forest • Prey: Spiny rats, other rodents and possums • Reproduction: Oviparous; 10–14 eggs.

Black-headed bushmaster *Lachesis melanocephala*
Range: Pacific coast of Costa Rica • Max. length: 2.0–2.4m • Venom: Pro- and anticoagulants, and haemorrhagins; snakebites rare but fatality rate high and rapid • Habitat: Tropical wet forest and rainforest • Prey: Rodents and possums • Reproduction: Oviparous; 9–16 eggs.

South American bushmaster *Lachesis muta*
Range: Amazon and Orinoco basins and Trinidad, and Atlantic coastal forest of Brazil • Max. length: 2.0–3.6m • Venom: Pro- and anticoagulants, and haemorrhagins, snakebites rare but fatality rate high and rapid • Habitat: Tropical rainforest and riverine gallery forest • Prey: Spiny rats, other rodents and possums • Reproduction: Oviparous; 7–20 eggs

Chocoan Bushmaster *Lachesis acrochorda*
Range: Panama, Colombia and Ecuador. Max. length: 2.0-2.3m. Venom: pro- and anticoagulants, and haemorrhagins Habitat: Tropical wet forests • Prey: Rodents • Reproduction: Oviparous.

Similar species: Terciopelo (*Bothrops asper*) and boa constrictor (*Boa constrictor*).

Below: *The Black-headed bushmaster (*Lachesis melanocephala*) is confined to the Pacific versant of Costa Rica.*

North American Rattlesnakes

Rattlesnakes evolved on the plains of North America and developed rattles to warn off potential dangers, the sound of a rattle being an excellent way to deter a herd of heavy-footed and short-sighted buffalo or a potential predator. All but two species of the 32 species of rattlesnakes are placed in the genus *Crotalus*, represented in every mainland American country except Panama, Ecuador and Chile, and also on a number of islands. They inhabit deserts, woodlands, grasslands, mountains, swamps and islands but not rainforests. Some rattlesnakes are incredibly rare such as the Autlan rattlesnake (*C. lannomi*), which is known from only a single specimen from Jalisco, Mexico. In selecting species I have chosen to represent all the habitats in which rattlesnakes occur. The common effects of rattlesnake bites are local swelling, bleeding, dangerous haemostatic disturbances, including a drop in platelet numbers, incoagulable blood, stroke, secondary renal failure and later, necrosis of tissue which may lead to amputation.

Western diamondback rattlesnake *Crotalus atrox*

The archetypal rattlesnake of cowboy movies, the western diamondback is a widespread species. A true generalist, in both habitat and prey preferences, it usually occurs in areas inhabited by several other species from which it may be distinguished by its banded black and white tail resulting in the local name of 'coontail rattlesnake'. In general coloration this can be an extremely variable snake, with specimens ranging from grey to yellow to red. It is a major cause of snakebite in southwestern USA.

Range: Southwestern USA, Arkansas to California, and northern Mexico • Max. length: 1.2–1.8m • Venom: Procoagulants and haemorrhagins; snakebites common with fatalities recorded annually • Habitat: Forest to grassland and desert • Prey: Rodents, rabbits and lizards • Reproduction: Viviparous; 4–61 neonates • Similar species: Black-tailed rattlesnake (*C. molossus*) and Mojave rattlesnake (*C. scutulatus*).

Eastern diamondback rattlesnake *Crotalus adamanteus*

The eastern diamondback achieves the greatest length of any rattlesnake, over two metres. A huge snake with a long strike, it is much feared, and this has led to its being also actively persecuted. It is also threatened by habitat destruction. Not afraid to swim, the eastern diamondback has colonized many of the islands and keys that dot the southern Florida coastline. Grey and yellow with black diamond markings and a black stripe starting behind the eyes, it injects huge amounts of venom and death is highly likely without treatment. Eastern diamondback rattlers are known to travel considerable distances during a single season and are able to navigate back to their overwintering site. Huge amounts of antivenom are required to treat their bites.

Range: Southeastern USA, Carolinas to Florida and Louisiana • Max. length: 1.8–2.4m • Venom: Pro- and anticoagulants, haemorrhagins, myotoxins; snakebites are serious and often fatal • Habitat: Upland pineland, palmetto scrub and coastal habitats • Prey: Birds, rodents and rabbits • Reproduction: Viviparous; 4–32 neonates.

Timber rattlesnake *Crotalus horridus*

The timber rattler is widespread in the USA, but the last Canadian specimen was killed near Niagara in 1941. It is one of the most arboreal rattlesnakes, often climbing into trees to hunt. Two subspecies have been recognized in the past, the northern, darker, timber rattler of the wooded slopes, and the lighter, southern canebrake rattler, so named for its liking for

canefields, which was characterized by an orange stripe down the back. Today, three distinct populations are recognized, northern, southern and eastern, but subspecies names are no longer allocated. There is also considerable geographical variation in venom composition. In the Appalachians the timber rattler is popular with the snake-handling congregation of the Church of Lord Jesus, who take the biblical quotation 'They shall take up serpents' [Mark XVI:18] literally. The timber rattlesnake was also the symbol of defiance chosen by the new nation of the United States against Great Britain in the War of Independence. The flag depicted a rattlesnake and the slogan 'Don't tread on me', but since those days the species has been actively persecuted, to virtual extinction in some areas.

Range: USA, New England to Texas (except Florida peninsula), exterminated from Canada by 1941 • Max. length: 1.0–1.8m • Venom: Procoagulants, haemorrhagins, myotoxins, serious snakebites, occasionally fatal • Habitat: Northern: rocky wooded slopes; southern: flatwoods, flood plains and gaps in cane fields • Prey: Rodents and birds • Reproduction: Viviparous; 3–16 neonates.

Opposite left: *The Western diamondback rattlesnake (*Crotalus atrox*) is distinguished from other species by its black and white 'coon-tail'.*

Opposite right: *The Timber rattlesnake (*Crotalus horridus*) was probably the first species encountered by colonists from The Mayflower but it is now highly endangered throughout its range.*

Above: *The Eastern diamondback rattlesnake (*Crotalus adamanteus*), a US endemic, is the largest species of rattlesnake.*

"My most serious snakebite was received from a one meter female 'canebrake', in a group that I was feeding with thawed out rats. Using her heat-sensitive pits she maneuvered into a position to strike me in the wrist. Defensive snakebites may be dry bites (no venom) or they may inject a sub-lethal dose of venom but prey-taking bites are intended to kill, and this one almost did. In the next five minutes I secured the snakes and locked the cage, collected the antivenom from the fridge and just managed to raise the alarm before the venom rendered me unconscious. Upon regaining consciousness I discovered I had lost my vision and virtually the ability to speak but I became incredibly sensitive to sound. In the ambulance I experienced an electrifying surge of venom and knew if I allowed myself to became unconscious again I might never wake up. Ironically I keep myself conscious for the twenty minute, high-speed, police escorted journey to hospital by trying to remember the names of all the rattlesnake species, in Latin. I usually get to around 29 easily and then have to think hard. I was hospitalized for nine days, lost the feeling in my arm for several weeks and almost lost my life.

Sidewinder rattlesnake *Crotalus cerastes*

The American equivalent of the Saharan sand viper or the Namib sidewinding viper, there are three subspecies of sidewinder rattlesnake inhabiting the Mojave and Sonoran deserts. The raised horns over the eyes serve to distinguish this small sand-coloured rattlesnake from any other rattlesnake. Inhabiting shifting sands, which it traverses using the characteristic motion that gives it its name leaving 'J' marks in the sand, or sheltering at the base of creosote bushes, this is a true desert rattlesnake that can inhabit regions with conditions too extreme for other species.

Range: Southwestern USA, California, Nevada and Arizona, and northwest Mexico • Max. length: 0.5–0.8m • Venom: Procoagulants, local swelling, pain and bleeding; no fatalites recorded • Habitat: Desert and desert scrub • Prey: Rodents, lizards and birds • Reproduction: Viviparous; 5–18 neonates • Similar species: Saharan sand viper (*Cerastes cerastes*) and Namib sidewinding viper (*Bitis peringueyi*).

Ridge-nosed rattlesnake *Crotalus willardi*

There are five subspecies of ridge-nosed rattlesnakes distributed southwards through the Sierra Madre Occidental from Southwestern USA into Mexico. They are known as 'ridge-nosed' because they exhibit a pronounced 'canthus rostralis', the line which separates the top from the side of the head and runs from snout tip to eye. Coloration varies from red-brown to grey and usually there is a series of light stripes running through the nostril to the angle of the jaw and across the lip scales. These are not the only small rattlesnakes inhabiting the mountains sometimes referred to as the Sky Islands (since they are frequently isolated peaks rising from the desert 'sea'). The ridge-nosed rattlers should not be confused with any other small montane US-Mexican rattlesnakes.

Range: Southwestern USA, California and Arizona, and northwestern Mexico • Max. length: 0.5–0.6m • Venom: Probable procoagulants, otherwise unknown • Habitat: Woodland and forest in canyons and on sheltered rocky slopes • Prey: Rodents, birds, lizards and centipedes • Reproduction: Viviparous; 4–8 neonates.

Santa Catalina rattlesnake *Crotalus catalinensis*

Endemic to Santa Catalina, this is a true desert island rattlesnake that has evolved to survive under extreme conditions with the loss of the rattle an important adaptation. In the absence of large, heavy-footed mammals from the small desert island, the snake does not require a rattle as a warning device. In the absence of any small mammalian prey it has adapted to feed on birds that roost in the bushes and has long fangs to penetrate their plumage. The presence of a rattle could be a distinct disadvantage when climbing to capture alert prey so the rattlesnake has become rattle-less. Each time it sheds its skin it fails to retain the final section and develop any sort of a rattle. Therefore it possesses only a terminal button which is incapable of making any sound even though the Santa Catalina rattlesnake does vibrate its tail when disturbed. Other islands in the Sea of Cortez are inhabited by rattlesnakes in the process of losing their rattles. An example is Isla San Esteban inhabited by the almost rattle-less San Esteban rattlesnake (*C. molossus estebanensis*).

Range: Isla Santa Catalina, off Baja California, in the Sea of Cortez, Mexico • Max. length: 0.5–0.7m • Venom: Not known • Habitat: Desert island cactus scrub • Prey: Birds and lizards • Reproduction: Viviparous • Similar species: Red diamond rattlesnake (*C. ruber*), which also has a rattle-less subspecies.

"During filming of 'Silent Rattler' in California we visited San Esteban, where we found a single rattlesnake, and Santa Catalina, where we found three: two females in bushes a metre off the ground and a male on the ground at night. The conditions on these desert islands in the Sea of Cortez were so extreme that we, like the rattlesnakes, soon became nocturnal.

Santa Catalina, Baja California, Mexico.

Opposite: *A trail of J-markings in loose southwestern desert sand indicates the passing of a Sidewinder rattlesnake (*Crotalus cerastes*), the only rattler with horns over the eyes.*

Above: *Living on an island with no heavy-footed mammals and feeding on roosting birds, the Santa Catalina rattlesnake (*Crotalus catalinensis*) has lost its rattle.*

Below: *The little Ridge-nosed rattlesnake (*Crotalus willardi*) is a protected montane species from the southwest.*

South American Rattlesnakes

The South American rattlesnake was previously known as the Neotropical rattlesnake because it contained up to 14 subspecies distributed from Mexico to Argentina. Two subspecies, the Aruba Island rattlesnake (*C. d. unicolor*) and Uracoan rattlesnake (*C. d. vegrandis*) from northern Venezuela, which were treated as full species, are again relegated as subspecies and recent analysis of the complex has led to the removal of all Middle American subspecies. The term South American rattlesnake is now confined to 11 subspecies, occurring to the east of the Andes. No rattlesnakes occur in Panama, Ecuador or Chile.

South American rattlesnake *Crotalus durissus*

Rattlesnakes are widespread in South America, with a broadly contiguous range from northeastern Brazil to Argentina and Uruguay. Further north the distribution is more fragmented with populations existing on isolated savannas in Venezuela and Colombia; along the Guianan coast; in the Rupununi savannas of Guyana/northern Brazil/southeastern Venezuela, and even in savanna pockets within the Amazon. Common characteristics include a dorsal rhomboid patterning and a pair of dark stripes extending from the top of the head onto the upper body'.

Large specimens possess a raised ridge along the spine of the anterior body. Stoutly built, they achieve proportions rivaled only by the eastern diamondback of North America, although populations from isolated savannas, and the Aruba Island rattlesnake, achieve far smaller sizes. South American rattlesnakes exhibit a wide geographical range of coloration and patterning and even venom composition. Bites from southern Brazilian rattlesnakes are mainly neurotoxic, similar to elapids, with minor swelling and few haemotoxic symptoms. Bites from northern Brazilian snakes are mainly haemotoxic and cytotoxic, with gross swelling, haemostatic disturbances and no neurotoxic symptoms, like bites from lanceheads (*Bothrops*). This has led to doctors trained in the south of Brazil failing to recognize rattlesnake bite symptoms in northern Brazil, causing at least one death due to improper treatment. This is a high-striking rattlesnake with an alarming flanking strike. Bites from rattlesnakes are believed to be responsible for nine per cent of all serious snakebites in Latin America.

Range: Colombia to Argentina, east of the Andes • Max. length: 1.0–2.0m • Venom: High geographic variability; Southern populations: presynaptic neurotoxins, only slightly haemotoxic or cytotoxic; Northern populations: procoagulants, haemorrhagins and cytotoxins, causing bleeding, swelling and necrosis, no neurotoxins; many snakebites with many fatalities • Habitat: Savanna and savanna woodland to semi-desert • Prey: Rodents, birds and lizards • Reproduction: Viviparous; 9–40+ neonates • Similar species: Middle American rattlesnake (*C. simus*) from Mexico to Costa Rica.

"I was bitten while photographing a small South American rattlesnake on a remote ecological station, at 18:30, 30 minutes after our last radio-check with our office in Boa Vista. There would be no communications until the next morning. The expedition nurse monitored the bite, measuring the swelling and conducting hourly observations while I noted the pain and discomfort. By early morning the swelling was worse and we decided to administer antivenom, 50 mls injected in 50 mls of normal saline, in a drip at a controlled rate of infusion. I immediately experienced difficulty in swallowing and developed a metallic taste and gum-boils. I also went blind, my vision being totally white. The drip was stopped, adrenaline was administered and slowly my vision recovered. The drip was restarted at a slower rate. At 06:00 our base was alerted and a medi-vac plane was sent. My journey also involved a jeep, a canoe and a 4x4. In the aircraft I was given further adrenaline for urticaria. I declined further antivenom at the Boa Vista hospital and made a rapid recovery. **Ilha Maracá, Roraima, Brazil.**

Left: *South American rattlesnake (*Crotalus durissus*) usually have dark neck strips and a raised dorsal ridge.*

Opposite top: *The swamp-dwelling Massasauga (*Sistrurus catenatus*) hibernates in crayfish burrows.*

Opposite bottom: *Pigmy rattlesnakes (*Sistrurus miliarius*) have such as small rattle it may be almost inaudible.*

Pigmy Rattlesnakes

The two small rattlesnakes of genus *Sistrurus* are separated from the typical rattlesnakes of genus *Crotalus* by the structure of the male's hemipenes (copulatory organs). Pigmy rattlesnakes also exhibit the enlarged regular head scutes more usually associated with colubrids and elapids, whereas the typical rattlesnakes demonstrate at least some degree of head-scale fragmentation. However, amongst typical rattlesnakes there are species with almost every stage of fragmentation from entire scutes, as per *Sisturus*, to complete fragmentation in the diamondbacks, so this characteristic is not as useful as formerly thought.

Pigmy rattlesnake *Sistrurus miliarius*

The three subspecies of pigmy rattlesnakes are only found in a variety of habitats but rarely far from water. In most specimens an orange vertebral stripe down the back contrasts with the dark blotches over a light grey background. The tiny rattle is almost inaudible and warnings from pigmy rattlesnakes may go unheeded and lead to bites that could cause serious symptoms in children.
Range: Eastern and central USA • Max. length: 0.5–0.8m • Venom: Haemorrhagins, local pain and swelling likely, snakebites recorded but no fatalities • Habitat: Pinewoods, palmetto scrub and grasslands • Prey: Lizards, mice, frogs and even grasshoppers • Reproduction: Viviparous; 2–18 neonates.

Massasauga *Sistrurus catenatus*

Also known as the swamp rattlesnake, the massassauga has a preference for damp habitats and even those populations in desert regions are localized around the water sources. *Massasauga* derives from a Native Canadian tribe, the Mississauga, meaning 'great river mouth', a reference to the swampy home environment of both the Native Indians and rattlesnakes. Massasaugas are usually grey but northern specimens may be melanistic (black) with little patterning. Being black means basking is more efficient in cooler weather. In winter massasaugas hibernate in old stumps, rocky fissures, or wet crayfish burrows, at depths below the frost line. Comprising of three subspecies, the massasauga is a protected species.
Range: Southeastern Canada, through midwest USA to Mexican border. • Max. length: 0.5–0.9m • Venom: Procoagulants, haemorrhagins and myotoxins; fatal snakebites have been documented.• Habitat: Swamps to prairies, limestone hillside to desert grassland • Prey: Lizards, mice, snakes, even centipedes • Reproduction: Viviparous; 5–12 neonates • Similar species: The Mexican pigmy rattlesnake (*Crotalus ravus*) was, until recently, considered the third *Sistrurus* species, but research suggests it may be more closely related the other small montane Mexican rattlesnakes and it has been moved to the rattlesnake genus *Crotalus*.

Venomous Snakes of the Caribbean

The West Indies consists of hundreds of large and small islands. The region is home to almost 700 species of amphibians and reptiles, although dangerous venomous snakes are found on only a few islands and endemic venomous snakes on only five islands. Several islands off the north coast of S. America are inhabited by South American species.

Roatán coralsnake *Micrurus ruatanus*
Babaspul *Micrurus nigricinctus babaspul*

The Roatán coralsnake is an endemic bi-coloured (black/red/black) monadal coralsnake. Locals do not consider it dangerous, based on the belief that to be venomous, a snake must feed on toads and there are no toads on Roatán! The tricolour monadal Babaspul (Creole for 'barber's pole', a reference to its pattern) is endemic to Great Corn Island, Nicaragua. Both coralsnakes are endangered and may even be extinct.

Range: Roatán, Islas de la Bahia off Honduras; Great Corn Island, off Nicaragua • Max. length: 0.5–0.6m • Venom: Probable postsynaptic neurotoxic and myotoxic; no snakebites recorded • Habitat: Tropical moist forest • Prey: small lizards and snakes • Reproduction: Oviparous, otherwise unknown.

Aruba Island rattlesnake *Crotalus durissus unicolor*

In the island's arid, cactus and thorn-scrub interior there is an endemic rattlesnake subspecies. Patterning consists of pastel yellows, mauves or pinks, the usual bold dorsal diamond rattlesnake patterning being reduced or absent. This endangered rattlesnake is the subject of a conservation and captive breeding program between US and European zoos.

Range: Aruba, 25km north of Venezuela • Max. length: 0.8–1.0m • Venom: Insufficiently known, likely to be coagulants and haemorrhagins; no snakebite reports • Habitat: Arid thorn-scrub • Prey: Whiptail lizards • Reproduction: Oviparous; 2–14 neonates.

St Lucia lancehead *Bothrops caribbaeus*
Martinique lancehead *Bothrops lanceolatus*

The volcanic islands of St Lucia and Martinique were probably seeded with ancestral South American lanceheads carried north on rafts of vegetation from the mouth of the Orinoco River. Both species are highly variable in colouration and patterning. The venom of the semi-arboreal Martinique lancehead is used in homeopathy remedies.

Range: St Lucia and Martinique, Lesser Antilles • Max. length: 1.0-2.0m • Venom: Procoagulants, haemorrhagins and myotoxins, fatal arterial thrombosis a possible consequence • Habitat: Tropical forest and plantations • Prey: Rodents, birds and mongooses • Reproduction: Viviparous, 30-75 neonates • Similar species: The only other large snake on St Lucia is the nonvenomous clouded boa contrictor (*Boa constrictor nebulosa*). No similar species on Martinique.

Islands with mainland species:

The inshore, **Isla de Margarita** is inhabited by the Venezuelan coralsnake, Lansberg's hognosed pitviper and a population of neotropical rattlesnakes. Further northeast lie the **Islas los Testigos**, a group of five small islets with habitat similar to Aruba, and a population of South American rattlesnakes. **Trinidad** resembles a section of South America that has broken away, rather than a separate Caribbean island. It has jungles, swamps, a mountain range rising to over 900m, and two coralsnakes, the aggressive terciopelo, and the only island population of the forest-dwelling Amazonian bushmaster.

Left top: *The Martinique lancehead (*Bothrops lanceolatus*) is one of only two endemic Caribbean lancehead species.*

Left bottom: *The Aruba Island rattlesnake (*Crotalus durissus unicolor*) is endangered in the wild.*

South American Rear-fanged Colubrids

For many years it was thought that the only dangerous rear-fanged colubrid snakes were African treesnakes but in recent years a number of serious snakebites have occurred from species previously though harmless or inoffensive. Some of these bites occur to herpetologists and herpetoculturists, instead of agricultural workers or villagers as is typical of viper and elapid bites. This is probably because the bites often require some degree of provocation and tend to occur when the snake is being captured or handled, rather than simply encountered in the field. Although serious effects are rare, and deaths even rarer, it is important to realize that no antivenom exists for the treatment of these bites, regardless of how serious they may become.

Argentinian black-headed snake *Phalotris lemniscatus*

This little snake used to be known as *Elapomorphus bilineatus* and that was its name when a South American herpetologist told me how he nearly died from its bite. Over a 20-day period, following a prolonged bite between his fingers, he suffered burning pain, swelling, localized bleeding and a severe headache before progressing to bloodstained urine and extensive bleeding, both internally and externally, and finally renal failure. But for medical care he would have probably died from this snakebite. This is a small, slender snake with two orange longitudinal stripes on a darker background, yellow underbelly and a yellow neck stripe behind its small black head. It is semi-fossorial (burrowing) and rarely seen unless discovered beneath a rock. It looks harmless.

Range: Brazil, Uruguay and Argentina • Max. length: 0.3–0.7m • Venom: Coagulants, haemorrhagins, but little is known; few snakebites, but capable of causing fatalities • Habitat: Rocky hillsides and grassland • Prey: Amphisbaenians, lizards, small snakes, frogs and invertebrates • Reproduction: Oviparous; 1–8 eggs • Similar species: Other members of genus *Phalotris*.

Lichtenstein's green racer *Philodryas olfersii*

One of approximately twenty species of closely related, fast-moving semi-arboreal or terrestrial racers, Lichtenstein's green racer is believed responsible for many painful but otherwise non-serious snakebites throughout its range. Apart from pain victims suffered swelling and discolouration which may last for several days. A fatal bite to a child in Brazil is reported.

Range: Brazil, Uruguay and Argentina • Max. length: 1.0–1.5m • Venom: Possible haemorrhagins and anticoagulants; snakebites and a fatal bite to a child • Habitat: Rainforest • Prey: Amphibians, lizards, birds and small mammals • Reproduction: Oviparous; 7–8 eggs • Similar species: Other members of genus *Philodryas* and related genera such as *Alsophis* and *Dromophis*.

Above: *Lichtenstein's green racer (*Philodryas olfersii*) is an aggressive avian predator reputed to have killed a child.*

"I have collected specimens of both *P. olfersii* and the related *P. viridissimus* and found both species extremely aggressive and inclined to bite given the opportunity, even chewing my boot. **Ilha Maracá, Roraima, Brazil.**

Other genera which have caused pain and alarming reactions include the following:

- Short-tailed snakes, genus *Tachymenis*, six species from southern S. America, probably the only snakes in Chile capable of painful snakebites with one possible fatality reported.
- False jararacas, genus *Xenodon*, five species from C. and S. America. I suffered prolonged bleeding following the bite of *X. rabdocephalus* in Honduras and a possible fatality to an Ecuadorian *X. severus* is reported.
- Parrot snakes, genus *Leptophis*, eight species from Mexico to South America. I received a bite from a large specimen of *L. ahaetulla praestans* in Central America, which caused a wave of pain across the bitten hand and arm, followed by a wave of numbness which lasted for several hours.
- False water cobra, *Hydrodynastes gigas*, from Brazil had caused some unpleasant symptoms in zoo personnel but my mother was bitten without ill effect.

EURASIA

Eurasia is a huge region comprising all of Europe and the majority of temperate Asia, but determining its borders with Africa and Tropical Asia is problematic, especially as any such barriers do not follow national frontiers. The zoogeographic zone known as the Palearctic encompasses much of what is included here as Eurasia, but, strictly, the Palearctic also includes Africa, north of the Sahara, which I have included in the African chapter, and omits the Arabian peninsula, which I have included within this chapter as Asiatic rather than African. The Himalayas form a reasonable division between Eurasia/Palearctic and Tropical Asia/Oriental but the lines are more blurred in southern China. I placed southern China, Taiwan and the Ryukyu Islands in Tropical Asia and Manchuria, the Koreas and Japan in Eurasia. This seems to roughly fit with the Palearctic-Oriental divide in the literature.

Eurasia, as here defined, includes regions as diverse as Scandinavia and the British Isles, Saudi Arabia and UAE, Afghanistan and western Pakistan, before skirting north of the Himalayas to Siberia, Korea and Japan. Habitats within this region include temperate woodlands, forests and grasslands, arid deserts and other semi-desert habitats, and vast areas of territory inhospitable to cold-blooded reptiles. Despite the geographical scale of the Eurasian region it is a fairly depauperate region for venomous snakes, the true vipers forming the bulk of the species present with pitvipers, elapids and venomous colubrids being represented by few species.

Southestern Europe and Turkey are the centre of diversity for true vipers, while the Eurasian pitvipers are primarily found in Far Eastern Asia and the few cobras occur in the Middle East and Arabia.

Right: *Two male and one female Northern adders (*Vipera berus*) demonstrating the typical sexual dichromatism common in European vipers. This is the only venomous snake in Scandinavia and the United Kingdom.*

Eurasian True Vipers

The dominant venomous snakes of western Eurasia are the true vipers. Most European species are small but larger species inhabit Turkey and the Middle East. Many possess the characteristic viperine zig-zag pattern.

Adders and Vipers

The names adder and viper are interchangeable – they mean the same thing. The true vipers are primarily a Euro-African group with the largest European/Middle Eastern genus being *Vipera* with some 30 species (even after six species were removed to the related genera *Daboia* and *Macrovipera*).

Northern cross adder *Vipera berus*

Also known as the European or common adder, this is one of the most widespread terrestrial snake species in the world, being found from the Atlantic coast, in England, Wales and Scotland, across continental Europe, from Spain to within the Arctic Circle in Scandinavia, and eastwards to Siberian Russia and the Amur region on the Pacific coast. The populations on Sakhalin Island, north of Japan, and in the Caucasus, are treated as subspecies. In the UK habitat includes heathland and moorland, in Western Europe is may be found in cultivated land, in woodland or in rocky habitat, in Russia it inhabits wooded steppe. Males are usually light grey with a black zigzag while females tend towards brown with a dark brown zigzag but in cold regions all black, melanistic, females are common. I have seen some unusual adders ranging from yellowish to pale grey with light blue spots on the side. The juveniles are red-brown and for many years in the 19th century, they were thought to be a separate dwarf species. During the entire 20th century adders only caused 12 deaths in the UK, the last in 1975.

Range: British Isles and Scandinavia to the Caucasus and Sakhalin Island, far eastern Russia • Max. length: 0.6–0.7m (1.0m Scandinavia) • Venom: Procoagulants and haemorrhagins, many snakebites, few fatalities • Habitat: Heathland to woodland and wooded steppe • Prey: Small mammals and lizards • Reproduction: Viviparous, 3–20 neonates • Similar species: Basque cross adder (*V.seoanei*), from northern Spain, formerly a subspecies of *V.berus*.

"My first venomous snake bites occurred with adders. The first was received whilst giving a demonstration of the differences between adders and grass snakes at a large show. I had not planned to demonstrate the most obvious of differences so graphically. The bite was minor and I declined any anti-venom but accepted an anti-tetanus injection, but following only weeks after the 1975 fatality there was a great deal of excitement amongst the show organizers. The second bite was received whilst photographing a large female and I decided not to go to hospital, preferring to weather the bite and make notes.

Orsini's meadow viper *Vipera ursini*

The meadow viper is European's smallest and least dangerous viper. It is also the one with the most fragmented distribution, most of the western populations being confined to mountain valleys separated from one another by vast tracts of unsuitable habitat. Five subspecies are recognized, the western races resembling small versions of the cross adders. Several of these montane populations are now threatened by habitat destruction or grazing and may be facing extinction.

Range: France, Italy, Hungary, Romania and Balkan countries to Greece, extinct in Bulgaria • Max. length: 0.5–0.6m • Venom: Procoagulants and haemorrhagins, no snakebites known • Habitat: Lowland meadows and upland grasslands • Prey: Locusts, lizards and small mammals • Reproduction: Viviparous, 5–18 neonates • Similar species: Steppe viper (*V. renardi*), from southern Russia, formerly a subspecies of *V. ursini*.

Left: *The diminutive Orsini's meadow viper (*Vipera ursini*) inhabits isolated alpine meadows across Europe.*

European Nose-horned Vipers

The vipers of southern Europe exhibit upturned snouts that range from only the slight upturn of the asp viper (*V. aspis*) to a pronounced 'horn' in the sand viper (*V. ammodytes*). They tend to inhabit arid, sandy habitats and the nose-horn may be an adaptation for living in these habitats.

European sand viper *Vipera ammodytes*

Although not noted to be aggressive, the three subspecies of European sand viper are usually considered the most dangerous of all European mainland vipers, due to the toxicity of their venom, the length achieved by the species and the length of the fangs (13mm). Like the cross adder (*V. berus*), this is a sexually dichromatic species, the males being grey with dark brown to black zigzag markings, while the female is usually brown with a dark brown zigzag. Alternative names for the sand viper include nose-horned viper and long-nosed viper, reflecting the pronounced snout horn comprised of several small scales. It may be diurnal or nocturnal in activity.

Range: Austria and northern Italy to the Balkans and Greece, Corfu and Turkey • Max. length: 0.85–0.95m • Venom: Presynaptic neurotoxins, procoagulants and haemorrhagins; many snakebites • Habitat: Arid, rocky hillsides and scrubland • Prey: Small mammals and birds, some lizards • Reproduction: Viviparous; 4–15 neonates • Similar species: Transcaucasian sand adder (*V. transcaucasiana*), from Turkey and Georgia, formerly a subspecies of *V. ammodytes*, and asp viper (*V. aspis*) from France and Italy.

Lataste's viper *Vipera latastei*

Lataste's viper, alternatively known as the snub-nosed viper, is an Iberian species with two subspecies, one of which also occurs in northwest Africa. The dark zigzag patterning on a grey to brown background is similar to that of other European viper species, while the upturned snout tip is intermediate between the slight snout upturning of the asp viper (*V. aspis*) and the strongly upturned snout of the sand viper (*V. ammodytes*). Diurnally active, like most European vipers, Lataste's viper is reported to be irascible. There are anecdotal accounts of Lataste's vipers living behind the thick bark of cork-bark trees and biting people harvesting the bark. I spent time in Spain and Portugal searching for this species but failed to find it.

Range: Spain, Portugal, Morocco and Tunisia • Max. length: 0.6–0.7m • Venom: Probably procoagulants; snakebites occasional and no confirmed fatalities • Habitat: Dry, rocky hillsides, arid scrubland and coastal dunes • Prey: Small mammals, birds, lizards and large invertebrates • Reproduction: Viviparous; 2–13 neonates • Similar species: Atlas mountain viper (*V. monticola*), from the Atlas Mountains of Morocco, formerly a subspecies of *V. latastei*.

Top: *The southeast European sand viper (*Vipera ammodytes*) is Europe's most venomous snake.*

Above: *Lataste's viper (*Vipera latastei*) from the Iberian Peninsula, is reported to hide behind cork bark and bite harvesters.*

Middle Eastern Mountain Vipers

Turkey and the Middle East are home to many species of vipers, often with limited distributions and frequently threatened by habitat destruction, active persecution, over-collecting and probably war.

Armenian rock viper *Vipera raddei*

Also known as Radde's viper, this montane snake is found on rocky hillsides, below the snowline. It is now thought endangered, due to habitat destruction and over-collection. Two subspecies are recognized, the eastern-most form, from Kurdistan, exhibiting a high degree of variation in coloration and patterning. Snakes may be grey or brown with orange or brown dorsal markings consisting of a zigzag, a broad stripe or a series of separate spots, that are lighter than the ground colour, and rows of dark spots on the flanks.

Range: Eastern Turkey to northern Iraq and Iran • Max. length: 0.7–1.0m • Venom: Procoagulants and haemorrhagins; snakebites rare, one fatality reported • Habitat: Rocky hillsides with sparse woodland • Prey: Rodents, lizards and grasshoppers • Reproduction: Viviparous; 3–18 neonates • Similar species: Wagner's viper (*V. wagneri*), from eastern Turkey and Iran.

Ottoman viper *Vipera xanthina*

At first glance, the Ottoman, or coastal viper, looks like a giant, stout northern cross adder (*V. berus*) – at least that was my thought when I first saw captive specimens in the 1970s. Specimens may be overall grey or light brown with a zigzag or series of dark-edged blotches along the back and smaller dark spots on the flanks. This really is a large viper. The largest specimens are found on the small islands of the Aegean and are Europe's largest venomous snakes. Although the largest specimens are found almost at sea level, the species may be found to altitudes of 2000–2500m in eastern Turkey.

Range: Extreme northeastern Greece, western Turkey and Aegean islands • Max. length: 0.7–1.3m • Venom: Procoagulants and haemorrhagins; its large size suggests the capability of serious snakebites • Habitat: Humid, rocky and well vegetated habitats • Prey: Small mammals, birds and possibly lizards • Reproduction: Viviparous; 2–15 neonates • Similar species: Palestine viper (*Daboia palestinae*), from Syria, Lebanon, Jordan and Israel, formerly treated as a subspecies of *V. xanthina*.

Top: *The Armenian rock viper (*Vipera raddei*), a montane Middle Eastern species, is thought to be endangered.*

Left *The Ottoman viper (*Vipera xanthina*) of Turkey is Europe's largest venomous snake.*

Blunt-nosed Vipers

Formerly contained in the genus *Vipera*, the blunt-nosed vipers, genus *Macrovipera*, comprises two North African species, one widespread species, and one European species confined to the Cyclades Islands southeast of Greece.

Levant viper *Macrovipera lebetina*

The Levant or Lebetine viper is a wide-ranging species with three subspecies from Turkey to Pakistan and a fourth in North Africa while the nominate subspecies is confined to Cyprus. Highly variable, varying from blue-grey to sand, with zigzags, spots or crossbars, its egg-laying habits are in contrast to the majority of live-bearing Eurasian vipers. Some populations of levant vipers may also be viviparous. This dangerous snake causes fatal snakebites from Cyprus to Iraq.

Range: Turkey to Afghanistan and Pakistan; Cyprus, and Tunisia to Algeria • Max. length: 2.0–2.1m • Venom: Procoagulant and haemorrhagic; snakebites rare to common with some fatalities • Habitat: Steppe, rocky hillsides, vegetated ravines and cultivated land • Prey: Mammals, birds and lizards • Reproduction: Oviparous; 5–35 eggs (some populations are viviparous) • Similar species: Moorish viper (*M. mauritanica*) and desert viper (*M. deserti*) from North Africa.

Milos viper *Macrovipera schweizeri*

Once treated as a subspecies of the Levant viper, the Milos viper is Europe's most endangered snake species occurring on only four small Greek islands. It varies from light grey to orange or black, with or without faint dorsal crossbars. A bird specialist, it feeds on migrating passerine birds, although rodents may also be taken. Vipers either ambush birds near waterholes or climb into bushes to capture them at roost. Despite official protection, up to 10% of the Milos viper population are lost to snake smugglers or killed on the roads annually. Such losses are hard to sustain and active conservation programs have been initiated, including captive breeding and reintroduction.

Range: Milos and three neighbouring islands, western Cyclades, Greece • Max. length: 0.7–1.0m • Venom: Procoagulant and haemorrhagic, snakebites reported • Habitat: Rocky, well vegetated hillsides • Prey: Birds and introduced rodents • Reproduction: Oviparous, up to 10 eggs • Similar species: Levant viper (*M. lebetina*).

Below: *The numerous subspecies of Levant vipers (*Macrovipera lebetina*) cause many snakebites across their ranges.*

Palestine Vipers

Some authorities retain the Palestine viper in the Euro-Middle Eastern genus *Vipera* but there has been a trend to place it in the same genus as the Asian Russell's viper (*Daboia russeli*)

Palestine viper *Daboia palaestinae*

This snake is a significant cause of snakebite in Israel and Palestine because it is common in the highly populated lowlands, even though it announces its presence by hissing and elevating the forepart of its body. It is a nocturnal species that preys on a wide variety of small animals. The Palestine viper bears a resemblance to a small Russell's viper (*D. russelii*), the head and body markings being similar, especially in those specimens which have a pale orange ground colour with the red-brown dorsal zigzag broken into individual lozenges. However, for a long time this species has been linked more closely with the Ottoman viper (*V. xanthina*) from Turkey to the north.

Range: Western Syria, Lebanon, Palestine, Israel and northwest Jordan • Max. length: 0.7–1.3m • Venom: Procoagulants and haemorrhagins; many snakebites including fatalities. • Habitat: Lowland forest, coastal grasslands and rocky hillsides • Prey: Mammals, birds, some lizards • Reproduction: Oviparous; 7–22 eggs • Similar species: Russell's viper (*D. russelii*) from Tropical Asia.

Below: *The large and common Palestine viper (*Daboia palaestinae*) causes numerous serious snakebites.*

Opposite top: *The Arabian horned viper (*Cerastes gasperettii*) is a loose sand-dwelling desert specialist.*

Middle Eastern Desert Vipers

The deserts of Arabia and the Middle East are inhabited by a range of highly venomous, short and stout, sit-and-wait ambushers with keeled scales. The Arabian horned, carpet and sand vipers are closely related to species in North Africa while MacMahon's viper could be called the Asian sidewinder.

Arabian horned viper *Cerastes gasperettii*

The southwest Asian member of the North African genus *Cerastes* is almost identical to its close Saharan relative. It occupies similar habitats from Arabia to Israel and Iran, that is, stony desert and particularly scrubland around oases or dry wadis. Although capable of sidewinding, the horned viper avoids dunes of loose shifting sand. Patterning consists of faint light-brown crossbars on a pastel sandy background. Not all specimens bear horns over the eyes, but the horns, when present, are composed of a single pointed scale. Two subspecies are recognized.

Range: Israel to Sinai, Arabian Peninsula, Iraq and southwest Iran • **Max. length:** 0.6–0.8m • **Venom:** Procoagulants and haemorrhagins; snakebites rare, fatalities likely but rare • **Habitat:** Desert and semi-desert. • **Prey:** Rodents, lizards and birds • **Reproduction:** Oviparous; 8–20 eggs • **Similar species:** Sahara horned viper (*C. cerastes*) from North Africa and Sahara sand viper (*C. vipera*) which enters Israel from Africa and Sinai.

Bottom: Pseudocerastes persicus *is known as False horned viper because its 'horns' are comprised of short clumps of scales rather than individual elongate scales.*

False horned viper *Pseudocerastes persicus*

The generic name *Pseudocerastes* suggests a false-*Cerastes* and this species is known as a false horned viper because, although the vipers do appear to bear horns over the eyes, they are not long, single-scale horns, like *Cerastes*, but short clumps of small scales. The venom also contains neurotoxins rather than the haemotoxins of true horned vipers. Two subspecies are recognized, a widespread eastern race from the deserts of Iraq to Pakistan and Oman, and a subspecies, called Field's false horned viper, from Sinai, Israel, Jordan and northwest Saudi Arabia.

Range: Sinai to Oman, Iran, Afghanistan and Pakistan • Max. length: 0.4–0.7m • Venom: Presynaptic neurotoxins; snakebites rare, no fatalities reported • Habitat: Desert and semi-desert • Prey: Rodents, birds and lizards • Reproduction: Oviparous; 11–21 eggs • Similar species: Arabian horned viper (*Cerastes gasperetti*).

Painted carpet viper *Echis coloratus*

Burton's painted carpet viper is a common and serious snakebite risk in Israel, Sinai and the Arabia Peninsula. Pastel grey or brown, with a pattern of broad blotches or cross-bands, its defensive display, rubbing the scales of its concentrically curved body together, generates a sawing sound due to serrations on the scales. An audible warning, similar to a rattlesnake's rattle, it conserves valuable moisture that would be lost by hissing.

Range: Western Arabian Peninsula, Sinai and extreme northwest Africa • Max. length: 0.75m • Venom: Procoagulants, haemorrhagins and cytotoxins; snakebites are common, four fatalities reported • Habitat: Rocky slopes • Prey: Small mammals, frogs and large invertebrates • Reproduction: Oviparous; 6–10 eggs • Similar species: Newly described Oman carpet viper (*E. omanensis*) from Oman and UAE.

Above: *The Painted carpet or saw-scale viper (*Echis coloratus*) lays eggs, an atypical trait in vipers.*

MacMahon's viper *Eristicophis macmahonii*

This viper's discoverer, Capt. MacMahon, noted that it gave away its hiding place by continual loud hissing and that its skin was so thin that it is difficult to capture without injury. Inhabiting sandy deserts from Iran to Rajesthan, India, this is the 'Asian sidewinder'. It can also bury itself quickly in loose sand, like the horned vipers (*Cerastes*), shuffling its body until only its eyes are visible, allowing it to ambush passing mammals and lizards. The prehensile tail enables it to climb into low bushes to capture roosting birds.

Range: Southern Iran and Afghanistan to Pakistan and northwest India • Max. length: 0.7–0.9m • Venom: Presynaptic neurotoxins and procoagulants; five snakebites reported with two fatalities • Habitat: Sandy desert • Prey: Rodents, lizards and birds • Reproduction: Oviparous; up to 12 eggs • Similar species Horned or sand vipers (*Cerastes* spp.).

Eurasian Pitvipers

Pitvipers inhabit eastern Eurasia, from where they spread into Tropical Asia and the Americas.

Mamushis

The only pitvipers in temperate Asia and far eastern Europe are a group loosely termed *mamushis*, although strictly that name is reserved for Far Eastern Asian species. The mamushis are probably close to the ancestors of the American copperhead, and its kin in genus *Agkistrodon*, which reached the Americas via the Bering land-bridge between Russia and Alaska. Many of the mamushis are distributed over a wide area but at least one species, the Shedao pitviper (*Gloydius shedaoensis*) is confined to Shedao or Snake Island, off the coast of China, while the Tsushima mamushi (*G.tsushimaensis*) is endemic to Tsushima Island, off Japan.

Japanese and Chinese mamushis *Gloydius blomhoffi*

Known from one insular subspecies from Japan and southern Sakhalin Island, Russia, and three mainland subspecies from Korea and China. Patterning usually consists of dark blotches on a pale background but all black or striped specimens do occur. Localised variations in venom composition compound the threat posed by these dangerous snakes. In Japan the mamushi is associated with folklore and is the origin for the legendary, aggressive, flattened, spitting, hissing, jumping, venomous *Tsuchinoko*.

Range: Japan, China, the Koreas and eastern Russia • Max. length: 0.5–0.7m • Venom: Variable, pro- and anticoagulants, haemorrhagins; many snakebites, occasional fatalities • Habitat: Marshes, meadows and woodlands • Prey: Frogs, lizards and small mammals • Reproduction: Viviparous; 4–10 neonates • Similar species: Rock mamushi (*G. saxatilis*) from the Koreas and China.

Opposite bottom: *The desert-dwelling MacMahon's viper (*Eristicophis macmahonii*) is often termed 'Asian sidewinder'.*

Below: *The Siberian pitviper (Gloydius halys) is the only pitviper to enter Europe, but it does not occur west of the River Volga.*

Siberian pitviper *Gloydius halys*

Up to eight subspecies are recognized though the extensive range of the Siberian or Halys pitviper. The common name is misleading, this pitviper only enters southern Siberia and is more associated with steppe than snow. Although sometimes referred to as Europe's only pitviper, it does not occur west of the Volga delta, on the northern shores of the Caspian Sea, from where it ranges to the Pacific coast. The Siberian pitviper is grey or brown with lighter crossbars but patterning varies geographically.

Range: Azerbaijan, Kazakhstan, Iran and Uzbekistan to Mongolia and Manchuria, eastern China • Max. length: 0.6–0.75m • Venom: Pro- and anticoagulants, haemorrhagins; occasional snakebites, no fatalities reported • Habitat: Steppe, valleys, mountain slopes and woodlands • Prey: Rodents, birds and lizards • Reproduction: Viviparous; 3–10 neonates • Similar species: Central Asian pitviper (*G. intermedius*).

Eurasian Cobras

Cobras are associated with Asia and Africa with only three species in Eurasia, all confined to Arabia and the Middle East, although fossil cobras are known from Europe. Of the three species, one is Asian in origin, one is African and the third species is a Middle Eastern endemic.

Caspian cobra *Naja oxiana*

The Caspian cobra is the western-most representative of the India cobra complex. It is generally drab brown or grey with a pattern of body rings which are most evident in juveniles. In contract to its Indian, Thai or Chinese relatives, this cobra usually lacks any distinctive hood markings. The Caspian cobra population is under threat, numbers are in decline and conservation measures may be required to guarantee its survival. The extinct cobra (*N. romani*) was once distributed from the Ukraine to Austria and France so the threatened Caspian cobra may be a surviving link to the long extinct European cobras.

Range: Turkmenistan, Uzbekistan, Tadjikistan, Iran, Afghanistan Pakistan and northwest India • Max. length: 1.5-1.9m • Venom: Postsynaptic neurotoxin, snakebites with fatalities reported • Habitat: Rocky hillsides and scrubland • Prey: Small mammals, reptiles, birds and amphibians • Reproduction: Oviparous, 6-19 eggs.

Arabian cobra *Naja haje arabica*

The Arabian cobra is a subspecies of the widespread Egyptian cobra (*Naja haje*) of N.Africa. The Arabian cobra is found primarily in highlands of the southwestern Arabian Peninsula, above 1,500m, that receive 300mm of rainfall each year. Specimens collected in the lowlands have probably been transported there by floods. Brown or yellow in colour, often with a black head, neck and tail, this is the longest venomous snake in the Middle East.

Range: Southwest Saudi Arabia and Yemen • Max. length: 1.5–1.8m • Venom: Postsynaptic neurotoxin, common snakebites, probable fatalities • Habitat: Highland, well-watered habitats • Prey: Frogs and toads, domestic fowl and small mammals, even other snakes • Reproduction: Oviparous; clutch size not recorded.

Sinai desert cobra *Walterinnesia aegyptia*

The Sinai desert cobra or desert blacksnake is the most widespread Middle Eastern cobra, although it has a fragmented distribution. Shiny black or dark brown, the desert cobra is a secretive, nocturnal, primarily fossorial snake that rarely spreads a hood, preferring to hiss, form an 'S' shaped neck curvature and strike, often with mouth closed. The large stoney-desert dwelling spiny lizards (*Uromastyx*) form a large proportion of its diet. Although there are other black snakes in the Middle East, the desert cobra, can be identified by its lack of a loreal scale between the preocular and nasal scales, a typical elapid characteristic.

Range: Sinai, Arabian Peninsula, Palestine, Israel, Jordan, Lebanon, Syria, Turkey and Iraq • Max. length: 0.7–1.3m • Venom: Postsynaptic neurotoxins; snakebites with unconfirmed fatalities • Habitat: Stony desert and semi-desert • Prey: Lizards, small mammals, amphibians and birds • Reproduction: Oviparous; clutch size not reported • Similar species: Arabian burrowing asp (*Atractaspis microlepidota*) and harmless Jordanian whipsnake (*Coluber jugularis asianus*).

Opposite top: *The Caspian cobra (*Naja oxiana*) may be a link to the extinct European cobras.*

Opposite bottom: *The Arabian cobra (*Naja haje arabica*) is a subspecies of the North African Egyptian cobra which inhabits highland areas in southeast Saudi Arabia and the Yemen.*

Below: *Although related to the true cobras, the Sinai desert cobra (*Walterinnesia aegyptia*) rarely spreads a hood, preferring to form its body into an S-shape before lunging.*

Arabian Burrowing Asps

The burrowing asps or stiletto snakes are primarily an African phenomenon but one species occurs in the Arabia Peninsula.

Arabian and Ein Geddi Burrowing Asps *Atractaspis microlepidota* ssp.

The small-scaled burrowing asp of North Africa is represented in the Middle East by two subspecies, the Arabian burrowing asp in the Arabian Peninsula and Yemen, and the Ein Geddi or oasis burrowing asp in Sinai, Palestine and Jordan. A shiny black snake with a pointed head and an inferior mouth, the defensive posture of this snake consists of it arching its back so that its snout touches the ground, from which position it can rapidly strike sideways, flicking a fang out of the mouth to stab its aggressor. Bites are rapid and serious but poorly documented.

In Somalia the African subspecies is called 'seven steps' or 'father of ten minutes', both references to the belief that its venom kills very quickly. Deaths have occurred in six hours, quicker than snakebite fatalities caused by cobras. The fossorial habits of the burrowing asps means they are rarely encountered above ground and are only really a danger on nights with heavy rainfall, when barefooted villagers are bitten. Bites cause intense pain, affect blood flow to the heart and lead to collapse and potential respiratory failure.

Range: Sinai Egypt, Jordan, Israel and Arabian Peninsula • Max. length: 0.9m • Venom: Sarafotoxin affecting blood flow in coronary arteries; snakebites are rare but fatalities reported • Habitat: Semi-desert • Prey: Lizards, snakes and small mammals • Reproduction: Oviparous; clutch size unknown • Similar species: Sinai desert cobra (*Walterinnesia aegyptia*).

Eurasian Rear-fanged Colubrids

Eurasia has few rear-fanged colubrid snakes of medical importance. In Europe only the Montpellier snake can be considered any sort of a threat. In temperate Asia the main risk from rear-fanged colubrids comes from the keelbacks of genus *Rhabdophis* which contains approximately twenty species. One species, the Southeast Asian red-necked keelback (*Rhabdophis subminiatus*) has caused serious snakebites and another species, the yamakagashi, has caused fatalities.

Montpellier snake *Malpolon monspessulanus*

The Montpellier snake is the largest European snake and one of the most widespread, north and south of the Mediterranean and across the Middle East to the Black and Caspian Seas. Western and eastern subspecies are recognized. Although variably patterned, the large size and the presence of enlarged scales over the eye should confirm identification. Diurnal, alert and fast-moving, it prefers to escape but if cornered, it will defend itself vigorously. Montpellier snakes are not usually considered dangerous, the venom causing little more than localized swelling and temporary numbness and I suffered no effects from a bite I received.

Range: Mediterranean Europe, North Africa and Middle East • Max. length: 1.8–2.0m • Venom: Possibly neurotoxins; snakebites rare, minor symptoms and no fatalities • Habitat: Rocky scrubland and semi-desert fringes • Prey: Lizards, mammals, birds, snakes • Reproduction: Oviparous; 4–20 eggs • Similar species: Moila snake (*M. moilensis*) from North Africa.

"To be an efficient predator the Montpellier snake needs to bask early to raise its body to optimum temperature for hunting. To do so they frequently take advantage of the heat on 'black-top' roads and consequently many are run over by cars.

Out of 11 road-killed snakes, in two weeks in Spain and Portugal, I identified ten Montpellier snakes. Spain and Portugal.

Yamakagashi *Rhabdophis tigrinus*

Keelbacks are freshwater snakes with rough scales. They are inoffensive in appearance, and most are harmless, except the keelbacks in genus *Rhabdophis* which are potentially dangerous. The yamakagashi is a common Far Eastern keelback. Being rear-fanged it is difficult for it to bite unless handled and most bites are to boys and snake catchers who initially experience localized swelling and headache, followed by loss of consciousness. At least three cases have progressed to kidney failure and death.

Range: Far eastern Russia, the Koreas, Taiwan and Japan • Max. length: 0.6–0.7m • Venom: Procoagulants, possibly haemorrhagins; snakebites are common, at least three fatalities • Habitat: Water meadows, and watercourses • Prey: Frogs, toads, sometimes fish • Reproduction: Oviparous; 18–25 eggs.

Opposite: *The Arabian burrowing asps (*Atractaspis microlepidotus*) have caused human fatalities.*

"In the 1970s I had a number of watersnakes and keelbacks in my private snake collection and was always on the look out for species I had not kept before. I obtained a red-necked keelback, then *Natrix subminiatus*, from a reptile dealer and was trying to count its teeth, to determine its subspecific identification, when I noticed enlarged rear-fangs. I contacted the dealer to tell him I thought the snake venomous but he assured me it wasn't. The snake failed to feed and died soon afterwards so I preserved it. Shortly after that a medical student in London, who seemed to have purchased a specimen from the same imported batch as myself, was bitten and seriously envenomed. It was not long after that the genus *Rhabdophis* was resurrected to include a number of Asiatic rear-fanged keelbacks.

Below: *The Montpellier snake (*Malpolon monspessulanus*) is one of Europe's largest and most widespread snakes.*

AFRICA

Africa is a diverse continent of great arid sandy or rocky deserts, ancient rainforests, vast savannas, massive river deltas and swamps, huge valleys and towering mountain ranges. The north, including the Sahara and coastal mountain ranges, shows affinities with Europe and Arabia, while sub-Saharan Africa contains the truly African species. Great physical features: the Nile, Ethiopian Highlands, Mt Kilimanjaro, Rift Valley, Sahara, Kalahari and Namib Deserts, pose different environmental challenges leading to Africa being the most diverse continent for venomous snakes.

Africa, not Asia, is 'Cobra Central' since the hooded serpents show far greater diversity in Africa than in India or Southeast Asia. Africa is also the place where many dangerously venomous snakes can be easily confused with harmless species. In most parts of the world it is possible to tell if a snake is venomous by remembering a few simple rules but in Africa the rules often break down and harmless-looking snakes turn out to be dangerously venomous. The same could be said for Australia but in Australia almost everything is venomous, whereas in Africa, most snakes are nonvenomous, it is just difficult for a non-expert to tell the innocuous from the lethal. Africa is also home to the fastest snake on the planet, the black mamba, and the snake with the World's longest fangs, the gaboon viper. The only significant group missing is the pitvipers.

To the east, Madagascar and the Seychelles are inhabited only by nonvenomous or rear-fanged species but front-fanged snakes are found on a few islands. In the Gulf of Guinea the islands of Fernando Po, Sao Tome and Principe have green mambas and forest cobras. In the Indian Ocean, Socatra, off Somalia, has a carpet viper, while Zanzibar, Pemba and Mafia, off Tanzania, have forest cobras, Mozambique spitting cobras and stiletto snakes.

Right: *This large and impressive gaboon viper was photographed in southeast Cameroon which makes it the East African species,* Bitis gabonica, *rather than the West African species,* B. rhinoceros, *which does not occur east of the Togo Gap.*

African Cobras

African cobra species can be quite neatly separated into those that spit and those that don't. In any single location there is likely to be a spitter and a non-spitter and they are usually easy to tell apart.

African Non-spitting Cobras

Africa is home to at least five non-spitting cobras and that number is likely to increase because one species is currently undergoing revision. All exhibit a cosmopolitan vertebrate diet, are active by day or night, and can climb and swim well. They all cause fatal snakebites, with death resulting through respiratory paralysis.

Egyptian cobra *Naja haje*

The Egyptian cobra is the cobra of Cleopatra, the royal snake of the Pharaohs and a more likely instrument of her suicide than an 'asp', which would have caused a painful and unpleasant death. Egyptian cobras are large snakes that exhibit a fragmented distribution surrounding the Sahara with populations along the Mediterranean coast, across the Sahel south of the Sahara and throughout East Africa. The black Moroccan Atlas Mountains population is sometimes recognized as a separate subspecies, as is the southwest Arabian population (*see* pages 64–5). Egyptian cobras are generally brown, with a dark band across the throat, and sometimes speckled with darker pigment. The head and snout are rounded. They can raise one-third of their length vertically, and spread a broad, rounded hood, with little provocation.

Range: North Africa and Arabia • **Max. length:** 1.8–2.5m • **Venom:** Postsynaptic neurotoxin; snakebite fatalities recorded • **Habitat:** Savanna and dry woodland to semi-desert • **Prey:** Mammals, birds, toads and other snakes • **Reproduction:** Oviparous; 15–20 eggs.

> "I received a single fang snakebite from a medium sized Egyptian cobra and experienced the rapid onset of neurotoxic symptoms: ptosis (drooping eyelids), flaccid facial paralysis and breathing difficulties before I received antivenom. The effects were rapidly reversed by a combination of antivenom and neostigmine. I was discharged from hospital the next day.

Snouted cobra *Naja annulifera*

The snouted cobra was formerly considered the southern subspecies of Egyptian cobra but they are very different snakes, especially in appearance. The snouted cobra is heavy bodied, it spreads a longer, narrower hood and its scales are matt compared to the shiny scales of the Egyptian. The head of the

snouted cobra is also more angular. Specimens may be unicolour brown or patterned with broad dark and light brown bands.

Range: Tanzania to South Africa and Namibia and Angola • Max. length: 1.9–2.1m • Venom: Neurotoxin; snakebite fatalities recorded • Habitat: Savanna and bushveldt • Prey: Mammals, birds, toads and other snakes, even puff adders • Reproduction: Oviparous, 8-33 eggs • Similar species: Angolan cobra (*N. anchetiae*) from Southwest Africa was recently a subspecies.

Forest cobra *Naja melanoleuca*

The Forest cobra is an agile, diurnal species that climbs well and is more aquatic than other true cobras of genus *Naja*. The longest African cobra, it is the most feared, by locals, and respected, by herpetologists. Specimens from West African rainforests are glossy black with black edged white scales along the lips, hence the alternative name, Black-and-white-lipped cobra. Specimens from West African savannas have black and yellow bands, with the lighter bands heavily suffused with black. Those from East African savanna-woodlands and Natal coastal forests are matt brown with black tails. The species is under revision and is likely to be split into several species.

Range: West and central Africa and scattered populations in east and southeastern Africa • Max. length: 1.5–2.7m • Venom: Postsynaptic neurotoxin; snakebite fatalities recorded • Habitat: Rainforest and plantation, but in drier woodland habitats in the southeast • Prey: Amphibians, reptiles, mammals, birds and fish • Reproduction: Oviparous; 15–26 eggs.

"I consider forest cobras to be tricky and dangerous snakes. I caught a large black specimen that came down an oil palm in Cameroon but forest cobras also enter human dwellings and I have removed them from under beds and inside washrooms. Bold snakes, they will hood and advance on any threat. An aggressive 2.0m male forest cobra I kept in captivity caused more heart-stopping moments than a king cobra. Cameroon, South Africa, Zambia.

Cape cobra *Naja nivea*

The Cape cobra is the smallest and southern-most non-spitting true cobra in Africa but it has the most toxic venom. Specimens may be yellow, brown, black or speckled black on a light colour. Diurnally active, it is extremely agile, easily climbing *Acacia* trees to raid weaver-bird nests.

Range: Southwestern Africa from Namibia and Botswana to South Africa, as far as the Cape • Max. length: 1.5–1.7m • Venom: Postsynaptic neurotoxin; snakebite fatalities recorded • Habitat: Grassland and semi-desert • Prey: Birds, small mammals, amphibians and reptiles • Reproduction: Oviparous; 8–20 eggs.

Opposite left: *The snake charmer's choice in N. Africa, the Egyptian cobra (*Naja haje*) hoods with little provocation.*

Opposite right: *The Snouted cobra (*Naja annulifera*) of southern Africa was until recently a subspecies of the Egyptian cobra.*

Top: *Forest cobras (*Naja melanoleuca*) are very unpredictable and are much respected. This is a Zambian specimen.*

Above: *The Cape cobra (*Naja nivea*) possesses the most toxic venom of any African* Naja.

Rinkhals and other African Spitting Cobras

Spitting cobras can defend themselves from 8-10ft away by sending twin jets of highly cytotoxic venoms into victim's eyes. It causes intense pain, temporary blindness and corneal damage that may lead to permanent blindness. Venom on unbroken skin is harmless and even if it enters the mouth it is destroyed by saliva. Spitting is defensive; spitting cobras hunt like other cobras. Spitting cobra bites cause extensive necrosis similar to puff adders bites. There are eight species of spitting cobra in Africa but more are likely to be described.

Mozambique spitting cobra *Naja mossambica*

This small and common snake is known as M'fezi in Zululand. Primarily nocturnal, it often hunts in and around human dwellings. M'fezi inhabit termite mounds and rocky crevices, but under an old fridge in an outbuilding may serve just as well. It does not always hood before spitting. Bites result in disfiguring and tissue loss, but deaths are rare.

Range: Tanzania to Namibia and South Africa, KwaZulu and Transvaal • Max. length: 0.8–1.5m • Venom: Postsynaptic neurotoxin and cytotoxin; few fatalities but disfigured survivors • Habitat: Savanna and coastal woodland • Prey: Amphibians, lizards, rodents and snakes • Reproduction: Oviparous; 10–22 eggs.

"One M'fezi capture we made was of a large specimen hiding under a boulder on a game reserve. When we moved the rock it lay on its back and spat repeatedly. It was quite comical to see this snake lying upside down but still spraying venom at anyone within range. KwaZulu-Natal, South Africa.

Above: *Rinkhals (*Hemachatus haemachatus*) is unusual, it has keeled scales, it is a live-bearer and it shams death.*

Below: *The Mozambique spitting cobra (*Naja mossambica*) is a common cause of ophthalmic accidents.*

Opposite left: *The Black-necked spitting cobra (*Naja nigricollis*) is highly variable with both all black and zebra-patterned specimens.*

Black-necked spitting cobra *Naja nigricollis*

This snake is Africa's largest spitter, capable of spitting 10ft without raising its head or hooding. It occurs throughout tropical Africa, excluding rainforests. Those in the West and Central African savannas are black with red throat bands while a brown form occurs in East Africa. Two former southwest African subspecies are now treated as species and it is likely that further species will be described

Range: Sub-Saharan Africa from Senegal to Somalia, southwest to Angola • Max. length: 2.0–2.7m • Venom: Postsynaptic neurotoxin and cytotoxin; causes fatalities, survivors have severe wounds • Habitat: Savanna and semi-desert • Prey: Amphibians, lizards, rodents, birds and snakes. • Reproduction: Oviparous; 8–20 eggs • Similar species: Zebra spitting cobra (*N. woodi*) and southwestern black spitting cobra (*N. nigricincta*).

Red spitting cobra *Naja pallida*

The Red spitting cobra is an attractive snake, bright salmon red contrasting with a broad black throat band and subocular tear-drop markings. The newly described Nubian spitting cobra, which has more than one black throat band, was formerly considered a northern population of Red spitter.

Range: East Africa from Somalia to Tanzania • Max. length: 0.7–1.2m • Venom: Postsynaptic neurotoxin and cytotoxin, few fatalities but disfigured survivors • Habitat: Savanna and semi-desert. • Prey: Frogs, birds and rodents • Reproduction: Oviparous; 6–15 eggs • Similar species: Nubian spitting cobra (*N. nubiae*) from Ethiopia to Egypt and West African brown spitting cobra (*N. katiensis*) from Senegal to Cameroon.

"As I was photographing a juvenile red spitter the snake spat into my right eye. It was very painful but because I acted immediately, flushing my eye out with water for almost two hours, I suffered no lasting effects.

Rinkhals *Hemachatus haemachatus*

The Rinkhals is the only live-bearing elapid in Africa and the only keeled-scaled cobra. South African, with an isolated Zimbabwe highlands population, rinkhals may be black, brown to zebra-striped. Live-bearing is useful strategy for reptiles in cool climates where eggs would perish. Although rinkhals spit venom they also defend themselves using 'thanatosis' or death-feigning. The rinkhals rolls onto its back, body limp, tongue protruding from open mouth, cloaca gaping, as if dead. Sometimes it is difficult to photograph a rinkhals in any other position.

Range: South Africa and Zimbabwe • Max. length: 1.0–1.5m • Venom: Neurotoxin and cytotoxin, causes ophthalmic (eye) injuries; few snakebites, fewer fatalities • Habitat: Lowland and montane grassland • Prey: Frogs and toads, also rodents and birds • Reproduction: Viviparous; 20–30 neonates.

"When photographing a captive Rinkhal in hospital grounds, I was hit in the right eye by a jet of venom. I bagged the snake then returned to the hospital and washed my eye thoroughly. I was to conduct an interview on camera the next day and didn't want to have a blood-shot eye. Fortunately my quick actions prevented any damage and my eye recovered in time for the interview. **Johannesburg, South Africa.**

Below: *The Red spitting cobra (*Naja pallida*) can be bright salmon pink with a black throat band.*

Specialist Cobras

There are specialized African cobras that do not belong to the typical cobra genus *Naja*. These snakes are generally less common and much less well known than their highly visible relatives.

Gold's tree cobra *Pseudohaje goldii*

The tree cobras are arboreal forest dwellers never found in open country. Gold's tree cobra is a large, shiny black snake with a yellow underbelly, and a short head and disproportionately large eyes. Human snakebites are rare due to the arboreal and secretive nature of tree cobras but they are believed to possess extremely toxic venom and be very dangerous. Although thought to feed on amphibians the occasional capture of tree cobras in squirrel traps may suggest that mammals feature in their diets also. The hood of the tree cobra is not as broad as that of a typical cobra although it will rear up and hood if threatened. The tail tip ends in a sharp spike, like that of the equally arboreal forest cobra (*Naja melanoleuca*). This spike may assist in climbing. So poorly known are the tree cobras that we don't know if they are really diurnal or nocturnal. They may represent a link between cobras and mambas.

Below: *Gold's tree cobra (*Pseudohaje goldii*) is an elusive and rarely seen arboreal species with disproportionately large eyes.*

Range: Central Africa from Nigeria to Uganda • Max. length: 2.2–2.7m • Venom: Probably neurotoxins, nothing known; no snakebites reported but considered dangerous • Habitat: Rainforest • Prey: Amphibians • Reproduction: Oviparous; 10–20 eggs • Similar species: Black tree cobra (*P. nigra*) from West Africa and Blanding's treesnake (*Toxicodryas blandingi*) from West and central Africa.

Banded water cobra *Boulengerina annulata*

An orange-brown snake with bold black bands along its entire body, the banded water cobra is a powerful aquatic snake. A subspecies, Storm's water cobra, is endemic to Lake Tanganyika, the world's longest and second deepest lake. Storm's water cobra is only banded around the neck. Water cobras are adept divers, hunting fish up to 25m beneath the surface and remaining submerged for 20 minutes. They are a sort of freshwater 'seasnake' but are also agile on land, unlike seasnakes. The water cobra can also spread an impressive hood if threatened, even underwater. Actively hunting amongst submerged rocks, the water cobra is an efficient predator. Fish are killed and swallowed extremely quickly underwater. Although not threatening to humans a bite would be very

Below: *The Banded water cobra (*Boulengerina annulata*) is a highly adapted aquatic predator of cichlids and other fish.*

serious. Local people claim that you will be okay if you remain in the water but the instant you come out onto land you will drop dead! The people most at risk are gillnet fishermen who are used to pulling drowned water cobras from the nets. A half-drowned cobra is quite likely to deliver a fatal bite.

Range: Central Africa from Cameroon to Tanzania and Zambia • **Max. length:** 2.4–2.7m • **Venom:** Probably postsynaptic neurotoxins; no snakebites reported but considered dangerous • **Habitat:** Rivers and lakes • **Prey:** Fish • **Reproduction:** Oviparous; clutch size unknown • **Similar species:** Congo water cobra (*B. christyi*) from the lower River Congo and watersnakes (*Grayia* ssp.).

"Storm's water cobra may be endangered. In Lake Tanganyika its distribution is very localized. It was once found in considerable numbers along the occasional shoreline patches of large boulder-sized rocks where it overnighted before venturing into the lake in the morning to hunt. This is where villagers set their gillnets and the water cobra population appears to be in decline, possibly due to losses through drowning in the nets. We also found a dead water cobra with wounds that suggested it was killed by an African fish eagle. Lake Tanganyika, Zambia.

Below: The Burrowing cobra (Paranaja multifasciata) is an extremely rare and poorly known Central African rainforest species.

Burrowing cobra *Paranaja multifasciata*

The burrowing cobra is one of the least known of Africa's venomous snakes. At first impressions it doesn't look like a cobra, being small, green to brown with every scale tipped with black and presenting a speckled appearance. The head is short and fairly rounded with a black cap broken only by a pale crossbar on the nape. Only the two black stripes over the cream lips beneath the eye are reminiscent of a cobra, especially as the snake rarely hoods. Little is known of its habits, it is thought to be a burrower but it may be more terrestrial, living in leaf litter and moving on the surface. Three subspecies are recognized,

Range: Central Africa from Cameroon to Congo • **Max. length:** 0.5–0.8m • **Venom:** Probably neurotoxins, nothing known; no snakebites reported • **Habitat:** Rainforest and wooded savanna • **Prey:** Amphibians and/or snakes • **Reproduction:** Probably oviparous but otherwise unknown • **Similar species:** African garter snakes (*Elapsoidea* ssp.)

Other African Elapids

Although cobras are the dominant African elapids, there are many other elapids ranging from diminutive harlequin snakes to deadly mambas.

Shieldnose & African Coralsnake

Southern African relatives of the cobras, these species are distinguished from all other African elapids by their enlarged shield-like rostral scales. Nocturnal and secretive, they are usually only encountered crossing roads at night after rain.

Shieldnose snake *Aspidelaps scutatus*

The stout-bodied shield-nose snake uses its enlarged shield-like rostral scale for burrowing through leaf-litter, under rocks and through loose topsoil. Generally brown or grey, with pale bands or dark scale-tip speckling, it shelters in holes during the day, emerging after dark or rain to hunt. It will raise its body and slightly flatten its neck, before making an awkward lunging strike accompanied by a viper-like hiss. Three subspecies are recognized, all from southern Africa.

Range: Southern Africa from Namibia to Mozambique • Max. length: 0.6–0.7m • Venom: Postsynaptic neurotoxin; usually mild symptoms, a child fatality recorded • Habitat: Savanna and sandveld • Prey: Small mammals, amphibians, lizards and snakes • Reproduction: Oviparous; 4–14 eggs.

African coralsnake *Aspidelaps lubricus*

A close relative of the shieldnose, the African coralsnake is a small and variably patterned snake. The most attractive subspecies, for which the name coralsnake is most applicable, occurs in the Cape Province, South Africa. It is salmon-red with black bands or rings and a black chevron on the head. The two subspecies found further north are yellow-brown with faint cross-bands or even patternless. The African coralsnake is not closely related to the American coralsnakes. It can raise its body and spread a narrow hood. Hunting on cool nights, it searches for sleeping lizards.

Range: Southwestern Africa from Namibia to South Africa • Max. length: 0.5–0.8m • Venom: Postsynaptic neurotoxin; usually mild symptoms, child fatalities recorded • Habitat: Desert edges and sandveld • Prey: Lizards, snakes, rodents and reptile eggs • Reproduction: Oviparous; 3–11 eggs.

Left: *The African coralsnake (*Aspidelaps lubricus*) is not closely related to American corals, despite its name and patterning.*

Right: *The Shieldnose snake (*Aspidelaps scutatus*) has an enlarged rostral scale on its snout for burrowing in loose material in search of prey.*

African Garter and Harlequin Snakes

Although large dangerous cobras and vipers dominate the venomous African snake fauna there are also numerous small elapids throughout the continent. Up to ten species of African gartersnakes are definitely elapids and some of the larger specimens may cause worrying snakebites, but the two harlequin snakes are so small as to be totally innocuous and they may not even belong in the Elapidae.

Southern African gartersnake *Elapsoidea sundevalli*

North American gartersnakes are harmless fish and frog eating colubrid snakes, which are popular as pets, but Southern African gartersnakes are members of the cobra family, Elapidae – common names can be misleading, especially when applied to snakes from different continents. One of the most widespread of the ten known species is the southern African gartersnake, a banded snake that exhibits a fragmented distribution and is currently accepted to contain five subspecies. African gartersnakes are relatively small, only one subspecies of the southern African gartersnake exceeds 1.0m. They are smooth scaled, secretive nocturnal burrowers with short rounded heads, small eyes and pointed snouts for digging. The most striking characteristic of the group is that most, though not all, species or specimens, are banded, hence the common name. Rarely encountered, they are timid and disinclined to bite, defensive behaviour being limited to body flattening or thrashing if handled.

Range: Namibia to Botswana and Zimbabwe to Mozambique and South Africa • Max. length: 0.7–1.0m • Venom: Not known, localized pain and swelling only; no serious snakebites recorded • Habitat: Coastal grassland and arid savanna • Prey: Small reptiles, reptile eggs and small mammals • Reproduction: Oviparous; 8–10 eggs • Similar species: East African gartersnake (*E. loveridgei*).

Spotted harlequin snake *Homoroselaps lacteus*

Both the harlequin snakes are small and secretive, and their relationship to the other elapids is hard to understand. Harlequin snakes have traditionally been placed within the Elapidae because they possess fixed front-fangs but they have also been placed in the Atractaspididae and more recent DNA analysis seems to confirm they are not closely related to other elapids and may belong elsewhere. Only two species of harlequin snakes are known, both are confined to southern Africa. They are so small they are considered harmless to humans. As its name suggests, the spotted harlequin snake is brightly patterned with black and yellow bands and a row of dorsal orange spots. The second species, the striped harlequin snake (*H. dorsalis*) is even smaller, rarely exceeding 0.3m in length.

Range: South Africa, Cape to Transvaal • Max. length: 0.4–0.6m • Venom: Not known, localized pain and swelling only; no serious snakebites recorded • Habitat: Coastal savanna and semi-desert • Prey: Small snakes and legless lizards • Reproduction: Oviparous; 6–16 eggs.

Top: *The South African gartersnake (*Elapsoidea sundevalli*) is a front-fanged member of the Elapidae, unlike the familiar and harmless North American gartersnakes.*

Above: *The Spotted harlequin snake (*Homoroselaps lacteus*) may, or may not, be an elapid but it is too small to be a threat to man.*

Mambas

The mamba, the deadly treesnake of Africa, is one of the best-known names of any dangerous snake but few people realize that there are four species.

Black mamba *Dendroaspis polylepis*

The black mamba is probably the most feared snake in Africa and it's name is associated with rapid death throughout the world. A long, slender, extremely agile and fast-moving, diurnal treesnake, which is equally at home on the ground, it might be seen basking in the early morning sun at the top of a tree, a spot it may return to repeatedly day after day from its sleeping retreat before going hunting. It is an alert snake with excellent vision. Its head is long and often referred to as 'coffin-shaped' with a protruding supraocular scale over each eye that gives the impression of a scowl. Black mambas are gunmetal grey or even brown but never black, apart from the inside of their mouths. The black mamba moves smoothly through the trees and also over uneven ground with considerable grace, it flows like oil with the upper part of its body elevated from the ground. Its patternless grey body makes it difficult to follow with the eye, due to a lack of 'points of reference', and it appears to move faster than it actually is. Although probably the fastest snake in the world, reports that it can overtake a racehorse at full gallop are exaggerations. In the breeding season two male mambas may engage in combat for the attention of a female. They will entwine about one another and race around trying to force the other to the ground but will not bite each other.

Although black mambas will bite if cornered, in common with other snakes they prefer to avoid contact and will adopt a series of defensive strategies before resorting to attack. Initially a black mamba will freeze, hoping to avoid discovery, it may then try to flee, but if pursued it will turn, raise its body high off the ground, flatten its neck into a narrow hood and gape, flashing the black interior of its mouth. If all else fails it may press home with a series of rapid stab and release bites with serious consequences for the victim. Sometimes victims have been bitten by a mamba that was fleeing towards them and delivered a bite as it went past, often high on the body and not felt at the time. The fang marks only being discovered when the victim started to experience double vision and the shirt was removed. By this time the mamba is long gone.

There was a time when a black mamba bite meant almost certain death. A survey of snakebites in South Africa from 1957 to 1963 recorded over 900 venomous snakebites but only seven confirmed black mamba bites. From the 900 snakebites, 21 ended fatally including all seven black mamba bites – a 100 per cent fatality rate. Forty years later people do survive black mamba bites provided they reach hospital in time.

Below: *The Black mamba (*Dendroaspis polylepis*) does not inflict many snakebites but people wha are bitten may die very rapidly.*

The black mamba probably possesses highly toxic fast-acting venom because in becoming an agile hunter it has substituted physical strength and robustness for speed and agility, and since its prey consists of rodents and other mammals with sharp and potentially dangerous teeth and claws it requires venom which will subdue a victim before it has a chance to injure the mamba.

Range: East and southern Africa, isolated locations in West Africa • Max. length: 3.0–3.5m • Venom: Neurotoxin and cardiotoxin, yield 100–120mg, lethal human dose 10–15mg, rapidly fatal • Habitat: Savanna woodland and rocky hillsides • Prey: Rodents • Reproduction: Oviparous; 6–17 eggs • Similar species: Harmless black treesnakes (*Thrasops* spp.).

"An herpetologist working with me in the Transvaal was bitten through the snake-bag by a black mamba, a typical snake-catcher's accident. He was rushed to military hospital in 15 minutes. They saved his life with antivenom but the doctor told me that another five minutes and he might have died. I know of an anecdotal account where three snake catchers were killed by the same mamba, which bit them all through the sack. Transvaal, South Africa.

Below: *The Eastern green mamba (*Dendroaspis angusticeps*) is an inhabitant of deep, verdant, coastal bush.*

Eastern green mamba *Dendroaspis angusticeps*

Of the three species of green mambas the eastern green mamba is the best known but even within its range many green mamba sightings are not mambas. Several species of nonvenomous green bushsnakes, *Philothamnus*, occur in eastern and southern Africa and are more commonly encountered than the secretive mamba. Since the maximum length for the harmless snakes is 1.3m, anything larger is a green mamba. Green mambas are diurnal and highly arboreal, but they will descend to the ground and may be seen crossing roads. The green coloration allows them to blend in superbly with their natural habitat and makes them invisible to prey, predators and even experienced herpetologists. Snakebites are rare due to the secretive, arboreal nature of the green mambas and although the eastern green mamba is the most frequently encountered of the 'greens' it is the species with the least toxic venom and only one fatality recorded. All mambas, however, must be considered capable of causing death. 'Greens' are common in Mombasa, Kenya, St Lucia, Natal, South Africa.

Range: East Africa from Kenya to Mozambique and Natal, South Africa • Max. length: 1.5–2.3m • Venom: Neurotoxin; one child fatality recorded • Habitat: Coastal forest, moist savanna, thicket forest and gardens • Prey: Birds, rodents and bats • Reproduction: Oviparous; 10–17 eggs • Similar species: Western green mamba (*D. viridis*), Jameson's mamba (*D. jamesoni*) and harmless green bushsnakes (*Philothamnus* and *Hapsidophrys* spp.).

African Vipers

All vipers in Africa are true vipers, there are no pitvipers on the continent.

Night Adders

The six species of African night adders are considered primitive vipers. They are placed in their own subfamily, Causinae, separate from the true vipers, Viperinae. Active by day and night, they are extremely 'colubrine' in appearance, lacking the angular viperine head, and retaining the large head scutes and round pupils of the nonvenomous and mildly venomous family Colubridae. They also lay eggs, an uncommon trait for vipers. In common with typical vipers they possess folding front-fangs and full venom apparatus, although their venom is not highly toxic to man.

West African night adder *Causus maculatus*

This snake is brown with dark rhombic blotches and a dark 'V' on the nape, but its snout is more rounded than in some other species. It defends itself by puffing loudly, raising the forepart of its body and striking wildly, sometimes lifting its body off the ground. This species is one of the commonest causes of snakebite in West Africa, but victims usually only suffer fever, localized swelling, pain in the lymph nodes and in severe cases a little localized necrosis (tissue-death).

Range: Senegal to Ethiopia and Angola • Max. length: 0.6–0.7m • Venom: Not well known; many snakebites, no fatalities • Habitat: Most habitats from forest to semi-desert • Prey: Amphibians • Reproduction: Oviparous; 6–20 eggs • Similar species: Rhomic night adder (*C. rhombeatus*) and Harmless common egg-eater (*Dasypeltis scabra*).

"I encountered a pair of male night adders in combat over a female in the garden of our expedition house. When they saw me they broke off and tried to escape. I pursued a male and as I caught up with him he performed a 'back flip' and almost managed to bite me,

demonstrating their wild strike behaviour. My reactions were quicker than his; I avoided his strike and bagged him. Bonjongo, Cameroon.

Snouted night adder *Causus defilippi*

The tiny snouted night adder will put on a defensive display similar to that of its larger relatives. It will strike wildly but its bite causes less serious injuries than larger night adders since local necrosis (tissue-death) is absent. This species has an extremely pointed snout with the tip slightly upturned. Patterning of dark blotches and a dark nape 'V' are similar to other night adders.
Range: Southeastern Africa from Tanzania to KwaZulu-Natal, South Africa • Max. length: 0.3–0.4m • Venom: Not well known; many snakebites but no fatalities • Habitat: Dry savanna and thicket woodland • Prey: Amphibians • Reproduction: Oviparous; 3–9 eggs • Similar species: Rhombic night adder (*C. rhombeatus*) of East and southern Africa.

“A small snouted night adder was brought to my base camp by locals who believed it was a juvenile puff adder. Sometimes local people are not very good at identifying their native snake species. Lake Tanganyika, Zambia.

Carpet and Horned Vipers

The arid regions of Africa are inhabited by short, squat, keel-scaled sit-and-wait ambushers.

West African carpet viper *Echis ocellatus*

The West African carpet viper is probably the most dangerous snake in West Africa. It is very common in areas frequented by humans and their stock animals and unlike other carpet vipers it also enters woodland. It has serrated body scales like other carpet vipers and saw-scales a warning to deter interference. Carpet vipers also cause snakebites in north and east Africa, the Middle East and south Asia, causing over 10,000 deaths a year.
Range: West Africa, Mauritania to Cameroon • Max. length: 0.5–0.6m • Venom: Procoagulants, haemorrhagins; extensive bleeding, shock, renal failure and many fatalities • Habitat: Savanna and savanna woodland • Prey: Small mammals, lizards, amphibians, scorpions and centipedes • Reproduction: Oviparous, 6–20 eggs • Similar species: White-bellied carpet viper (*E. leucogaster*) from northwest Africa and common egg-eater (*Dasypeltis scabra*).

Saharan horned viper *Cerastes cerastes*

North African horned and sand vipers inhabit desert habitats, sidewinding across loose sand and shuffling down into an ambush position, leaving only the eyes and the horns visible. Saharan horned vipers ambush small mammals, lizards and even birds. The horn over the eye, a single elongate scale, is not present in all populations and 'hornless horned vipers' are common. The highly keeled scales of the body serve to collect early morning dew that can be lapped, the only available moisture, thereby providing the viper with fresh water. Although not as dangerous as the carpet viper some bites have proved life-threatening with at least one near fatality reported. Unlike the carpet viper this is a popular viper in captivity and there is a danger that keepers under-rate its potential.
Range: Sahara Desert • Max. length: 0.6–0.9m • Venom: Procoagulant; fatalities possible although most snakebites are minor • Habitat: Sandy and rocky desert, especially with sparse vegetation • Prey: Rodents, lizards and birds • Reproduction: Oviparous; 10–23 eggs • Similar species: Arabian horned viper (*C. gasperetti*), Sahara sand viper (*C. vipera*).

Opposite top: *The round pupils of the West African night adder (*Causus maculatus*) add to its harmless appearance.*

Opposite bottom: *The Snouted night adder (*Causus defilippi*) is sometimes mistaken for a juvenile puff adder.*

Top: *Saharan horned vipers (*Cerastes cerastes*) have caused some very serious snakebites.*

Above: *The West African carpet viper (*Echis ocellatus*) is extremely dangerous and causes many deaths.*

Bushvipers

A dozen species of bushvipers are found across central Africa, from the rainforests of West Africa to the mountains of East Africa. Some species exhibit extremely limited ranges, others are more widespread and new species are being discovered with regularity. Bushvipers are the African true viper equivalent of the bamboo pitvipers (*Trimeresurus*) of Asia and forest-pitvipers (*Bothriopsis*) and palm-pitvipers (*Bothriechis*) of Latin America.

Variable bushviper *Atheris squamiger*

The most widespread and variably patterned of the bushvipers, this species may be green, turquoise, grey-green, yellow, orange or red. An agile, slender climber with a long prehensile tail, it inhabits the rainforests of Central Africa and is a sit-and-wait ambusher of small vertebrates. It drinks without moving, lapping water accumulating on its heavily keeled scales.

Range: Central Africa, from Nigeria to Uganda and Angola • Max. length: 0.7–0.8m • Venom: Procoagulants; most bites minor but at least one fatality • Habitat: Rainforest • Prey: Rodents, lizards, amphibians and snakes • Reproduction: Viviparous; 7–9 neonates • Similar species: Rough-scaled bushviper (*A. hispidus*) from Congo-Uganda border and western bushviper (*A. chlorechis*) from West Africa.

Great Lakes bushviper *Atheris nitschei*

The stout-bodied, black and green Great Lakes bushviper is highly agile, occurring in trees and bushes, at altitudes of 1,000-2,800m, and easily climbing papyrus, elephant grass or bamboo. Generally nocturnal, it is an adept ambusher of arboreal lizards but will also hunt on the ground. Juveniles often resort to caudal luring, slowly waving the tail tip to attract prey within strike range.

Range: East Africa, Uganda to central Lake Tanganyika on the western (Congolese) shore • Max. length: 0.6–0.7m • Venom: Procoagulants; few bites recorded • Habitat: Moist savanna and forest, swamps and lakesides • Prey: Rodents, lizards, especially chameleons, and amphibians • Reproduction: Viviparous; 4–13 • Similar species: Mount Rungwe bushviper (*A. rungweensis*) from southern Lake Tanganyika.

Specialist Vipers

There are several unique species of African vipers that are placed in their own monotypic genera, but that are considered closest to the arboreal African bushvipers.

Udzungwa worm-eating viper *Adenorhinos barbouri*

Also called the short-headed viper, this is one of Africa's least known snakes. Confined to the two Tanzanian mountain ranges up to 1800m, it is terrestrial in habit. A nocturnal, rainforest or bamboo-thicket, leaf-litter species, thought to eat earthworms, it is occasionally found in newly-cleared plots or at forest edges, but deforestation will cause its extinction.

Range: East Africa, only from Udzungwa and Ukinga Mountains, Tanzania • Max. length: 0.3–0.4m • Venom: Nothing is known; no snakebites reported • Habitat: Bamboo and montane forest • Prey: Soft-bodied invertebrates, earthworms and slugs, and small frogs • Reproduction: Oviparous; 10 eggs in three females • Similar species: Usambara bushviper (*A. ceratophorus*).

Opposite top: *The Variable bushviper (*Atheris squamiger*) occurs in many colours over a wide distribution.*

Opposite bottom: *The green and black Great Lakes bushviper (*Atheris nitschei*) is one of the most stunning bushvipers.*

Below: *The Lowland swamp viper (*Proatheris superciliaris*) possesses a deceptively dangerous venom.*

Lowland swamp viper *Proatheris superciliaris*

Also called the floodplain viper, because of its distribution along the Zambezi River, the lowland swamp viper exhibits fragmented head scales and a strongly blotched pattern. Once thought rare, it is sometimes found in large numbers although populations appear localized. Although related to the arboreal bush vipers, this is a terrestrial species that enters rodent burrows, eats the occupants and takes up residence.

Range: East Africa, Lake Malawi from Tanzania and Malawi to the River Zambezi, Mozambique • Max. length: 0.5–0.6m • Venom: Procoagulants; considered extremely dangerous and capable of causing renal failure • Habitat: Moist grassland, swamps and flood plains • Prey: Rodents and amphibians • Reproduction: Viviparous; 3–16 neonates.

Kenya mountain viper *Montatheris hindii*

The tiny Kenya mountain viper has the smallest distribution of any venomous African snake. Confined to Mt Kenya and Aberdare National Parks at altitudes of 2,700-3,800m, it is terrestrial, diurnal and sluggish. Vulnerable to birds-of-prey, habitat loss and human persecution, it shelters from predators, wind and rain inside tussock grass, only emerging on sunny days.

Range: East Africa, only on Aberdare Range and Mt Kenya • Max. length: 0.3–0.35m • Venom: Procoagulants, otherwise nothing known. • Habitat: Open montane moorland • Prey: Lizards and frogs • Reproduction: Viviparous; little known, 2–3 neonates recorded.

Big African Vipers

The genus *Bitis* contains five of the the largest and heaviest venomous snakes in Africa, but it also contains some of the smaller, more specialized desert and mountain vipers.

African Puff adder *Bitis arietans*

The most widespread African venomous snake, and the cause of many snakebites, the stout-bodied, broad-headed, puff adder is named for its loud, drawn out, angry hiss. If the warning is ignored it will launch a rapid and far reaching strike. A sit-and-wait ambusher found in most habitats from sea-level to 3,500m, bites from the chevron-marked puff adder cause excruciating painful and debilitating injuries, often resulting in loss of digits or limbs, though they are not as immediately life-threatening as cobra or mamba bites. Adult puff adders feed mostly on bulky rodents and possess highly cytotoxic venom to speed up digestion of the prey, but juveniles feed on lizards and their venom contains a neurotoxin, to prevent the prey escaping, which diminished as the snake matures. One adult adder even swallowed a young leopard tortoise. Adult puff adders are so large that they move using the undulating 'rectilinear motion' (caterpillar crawl), more associated with large boas and pythons. Puff adders are found throughout non-forested sub-Saharan Africa with isolated populations in Morocco and Arabia. The Somali population is treated as a separate subspecies.

Range: Sub-Saharan Africa • Max. length: 0.9–1.9m • Venom: Cytotoxins, possibly haemorrhagins and/or coagulants; many snakebites, few fatalities • Habitat: Most habitats, excluding rainforest and desert • Prey: Rodents, birds, reptiles and amphibians • Reproduction: Viviparous; 20–40 neonates, record 156 • Similar species: Ethiopian mountain viper (*B.parviocula*) known only from a handful of specimens.

Gaboon vipers *Bitis gabonica* and *Bitis rhinoceros*

Formerly a single species with two subspecies, the two populations now receive specific status. The widespread East African gaboon viper (*B. gabonica*) occurs in central and east Africa, with isolated populations in Tanzania, Mozambique and South Africa (Natal) while the West African gaboon viper (*B. rhinoceros*) inhabits the coastal rainforests of west Africa. The West African gaboon has raised horns on the snout which are much less pronounced in the East African gaboon which, instead, possesses a large tear-drop marking under each eye, which is smaller or absent in the West African gaboon. Both gaboons exhibit the geometrical 'Persian carpet pattern' of pastel blues, buffs and mauves, and large leaf-like head that render them virtually invisible in leaf-litter. Gaboon vipers hunt rats as sit-and-wait ambushers, like the puff adder, and may take larger prey including, ironically, mongooses. The consequences of a gaboon viper bite are also more serious than a puff adder bite

Below left: *The huge and widespread Puff adder (*Bitis arietans*) is found throughout Africa, excluding rainforest and deserts, and is a major cause of snakebite.*

Below right: *The East African gaboon viper (*Bitis gabonica*) has a black tear-drop marking under the eye but only slightly raised nasal horns.*

with a more directly life-threatening effect on the cardio-vascular system. Venom injected deeply with 2.5cm fangs, the longest known for any snake, may rapidly cause death.
Range: West Africa, Guinea to Togo; central and East Africa, Nigeria to Uganda to South Africa (KwaZulu-Natal) • Max. length: 1.2–2.0m • Venom: Haemorrhagins and coagulants, many snakebites, frequent fatalities • Habitat: Forest, woodland and savanna-woodland in south • Prey: Mammals and birds • Reproduction: Viviparous;16-30 neonates, record 60 • Similar species: Rhinoceros viper (*B. nasicornis*)

"One of the most rewarding things I have done was participate in the release of two female gaboon vipers in the protected coastal woodland in the St Lucia National Park. It was wonderful to watch the two snakes crawl away and literally disappear before my eyes in the leaf-litter of the woodland edge. The population in S.Africa is a species on the edge and although afforded total protection in the country the gaboon viper is still endangered, especially when it tries to haul its bulk across the busy coastal highway. **St Lucia, KwaZulu-Natal, South Africa.**

Rhinoceros viper *Bitis nasicornis*

The rhinoceros viper is a true rainforest species with a preference for damp habitats which earns it the alternative name of 'river jack'. More brightly patterned blue, yellow and maroon, than the gaboon viper and with a group of raised horn-like scales on the tip of its snout, the rhinoceros viper is equally invisible in the leaf-litter, the cryptic patterning breaking up its outline. Being a deep forest snake it comes into contact with humans less frequently than its three larger relatives.
Range: West and central Africa from Liberia to Uganda and Congo, but not the dry Togo Gap • Max. length: 0.9–1.2m • Venom: possibly haemorrhagic and/or coagulant; few snakebites • Habitat: Rainforest and riverine forest • Prey: Mammals, amphibians, possibly fish • Reproduction: Viviparous; 6–38 neonates • Similar species: Gaboon vipers (*B. gabonica* and *B. rhinoceros*).

The puff adder is an arid savanna and savanna woodland species, the gaboon vipers are found from dry woodland to rainforest habitats while the rhinoceros viper is a tropical riverine-forest species. Where the ranges overlap there have been isolated cases of hybrid specimens reported ie. between puff adder and gaboon viper and between gaboon viper and rhinoceros viper. No puff adder-rhinoceros viper hybrids are known.

Below left: *The West African gaboon viper (*Bitis rhinoceros*) has no black markings under the eye and pronounced nasal horns, especially in adults.*

Below right: *The Rhinoceros viper (*Bitis nasicornis*) is also known as the River jack, due to its preference for riverine forest habitats.*

Small African Vipers

Twelve of the species in genus *Bitis* are small species which either inhabit deserts or mountains. Several exhibit specializations for their lifestyles.

Horned adder *Bitis caudalis*

The widespread horned adder is often confused with other small southern African horned vipers. A short, stout snake, it burrows into the sand to ambush lizards.

Range: Angola, Namibia, Botswana and South Africa • Max. length: 0.4–0.5m • Venom: Presynaptic neurotoxin and myotoxin; snakebites cause pain, swelling and local necrosis • Habitat: Sand and scrub desert • Prey: Lizards, amphibians and small mammals • Reproduction: Viviparous; 4–19 neonates • Similar species: Many-horned adder (*B. cornuta*), Albany adder (*B. albanica*) and desert mountain adder (*B. xeropaga*) from Namibia and South Africa.

Berg adder *Bitis atropos*

This is a southern African upland snake although it occurs at sea-level in the temperate Cape Province. It has neurotoxic venom, unusual for an African viper, to deal with small, fast-moving lizards. Snakebites are rare.

Range: Drakensburg Ranges of South Africa, Swaziland, Lesotho and Zimbabwe/Mozambique border • Max. length: 0.4–0.6m • Venom: Neurotoxins; snakebites with localized symptoms, one fatality reported • Habitat: Cool, damp, montane or southern coastal heathland called fynbos • Prey: Birds, rodents, lizards and amphibians • Reproduction: Viviparous; 8–16 neonates.

Peringuey's adder *Bitis peringueyi*

Peringuey's adder is Africa's famous sidewinder and the only viper completely adapted to the inhospitable Namib Desert. Sidewinding allows it to move diagonally over loose, hot sand with minimal contact, leaving characteristic 'J'-shaped markings in the substrate. The eyes of this sandy-orange, hornless adder are positioned on top of its head like a flatfish, an adaptation for lying almost buried in the sand.

Range: Namib Desert from Angola to Namibia • Max. length: 0.25–0.3m • Venom: Not known; few snakebites, no serious symptoms • Habitat: True sand desert • Prey: Sand lizards and barking geckos • Reproduction: Viviparous; 4–10 neonates • Similar species: Namaqua dwarf adder (*B. schneideri*) occurs in semi-desert further south.

Top: *The Horned adder (Bitis caudalis) is one of several short desert horned or hornless vipers from southwestern Africa.*

Middle: *The Berg adder (*Bitis atropos*) is unusual for a viper, it has neurotoxic venom.*

Left: *Peringuey's adder (*Bitis peringueyi*) is the ultimate desert snake and is also known as the Namib sidewinder.*

North African Vipers

The vipers of the North African coast and mountains show closer relationships with the vipers of Europe and the Middle East than with their relatives in Sub-Saharan Africa. The climate is Mediterranean temperate and the habitat varies from sandy desert fringe to wooded hillsides.

Moorish viper *Macrovipera mauritanica*

This snake is the largest viper along the North African coast. It resembles a typical viper of Middle Eastern origin, with a dark zigzag running the length of its brown or red back. Its activity cycle is crepuscular (active at twilight) when it hunts as an ambusher. Days are spent under cover, in dense brush or burrows.
Range: Northern Morocco and Algeria • Max. length: 1.8m • Venom: Probably procoagulants and haemorrhagins; no bites recorded but large size indicates dangerous levels of venom

Above: *Although found in Africa, the Moorish viper (*Macrovipera mauritanica*) is more closely related to vipers in the eastern Mediterranean.*

• Habitat: Wooded steppe and rocky mountains • Prey: Small mammals • Reproduction: Oviparous; 18 eggs • Similar species: Desert viper (*M. deserti*) from Tunisia and Libya.

Atlas mountain viper *Vipera monticola*

The little Atlas mountain viper is poorly known. It resembles a European adder in general appearance but is smaller than many European species. The snout is slightly upturned but not to the degree of Lataste's viper of southwestern Europe and the nose-horned or sand viper of southeastern Europe. Its small size is probably related to its high altitude existence, between 2400–3000m, where it hunts during the day and shelters during the night (in contrast to the larger Moorish viper).
Range: Atlas Mountains, Morocco. • Max. length: 0.3–0.34m • Venom: Probably procoagulants and haemorrhagins; no bites known and of small size. • Habitat: Rocky slopes and valleys • Prey: Lizards • Reproduction: Viviparous; 2 neonates • Similar species: Lataste's viper (*V. latastei*) of Moroccan coast and Iberian peninsula.

African Burrowing Asps

Also called mole vipers or side-stabbing snakes, and formerly placed in families Viperidae and Colubridae, the 18 species of burrowing asps, and related snakes, are now allocated their own family, Atractaspididae. Confined to Africa, southwestern Arabia and Israel, most countries only contain a single species, although Congo has eight and Cameroon, nine species. Many species are rare and most are difficult to identify. They possess the longest fangs, relative to head length, of any snake and can bite without opening their mouths. The chin is retracted allowing the horizontal fang to protrude from the side of the mouth. A turn of the head and the fang stabs home. It is impossible to hold a burrowing asp behind the head without being bitten. The warning posture of neck arched and snout tipped into the ground is characteristic of the group. Although the frequent cause of snakebites, especially to snake catchers, only three species have a history of causing deaths.

Bibron's stiletto snake *Atractaspis bibronii*

The most familiar burrowing asp is Bibron's stiletto snake which, like its relatives, is small, drab grey, smooth scaled, with small eyes, a slightly pointed head for excavating and a sharp spine on its tail tip for forcing it forwards through the soil. The underside is white but generally not visible when the snake is encountered at night. The venom is not fully understood and there is no antivenom available.

Range: Kenya to Namibia and South Africa • Max. length: 0.6–0.7m • Venom: Possibly sarafotoxin affecting blood flow in coronary arteries; snakebites but no fatalities • Habitat: Savanna, semi-desert and dry woodland • Prey: Reptiles, frogs and small mammals • Reproduction: Oviparous; 6–7 eggs.

> "I received a triple bite to both hands from a stiletto snake on a road in the Transvaal. It was extremely painful and both hands became very swollen. The Cuban doctor in the hospital 90 minutes away was not experienced in snakebite treatment and planned to use cryotherapy on my hands, to pack them with ice. This dangerous technique is no longer considered a suitable treatment for snakebite since it may result in tissue-death and loss of fingers. I took my discharge from the hospital and had to suffer a long night of pain and hallucinations before the effects of the venom wore off. **Transvaal, South Africa.**

Natal blacksnake *Macrelaps microlepidotus*

The Natal blacksnake may constitute a potential threat to children. Related to the burrowing asps, it is thought to possess a similar sarafotoxin. Bites have occurred and victims have lost consciousness but otherwise little is known about the risk posed by this rare snake.

Range: Endemic to KwaZulu-Natal, South Africa. • Max. length: 0.6–0.9m • Venom: Possibly a sarafotoxin; few snakebites, no fatalities • Habitat: Coastal woodland • Prey: Rain frogs, legless lizards, other snakes and small mammals • Reproduction: Oviparous; 3–10 eggs • Similar species: Apart from other burrowing asps, any dark snakes, i.e. purple-glossed snakes (*Ambylodipsas* spp.), unicolour gartersnakes (*Elapsoidea* spp.).

Top left: *Bibron's stiletto snake (*Atractaspis bibroni*) has a warning display which involves arching it back and tipping its snout into the earth as seen here.*

Left: *The Natal blacksnake (*Macrelaps microlepidotus*) is a very rare species confined to KwaZulu Natal, South Africa.*

African Rear-fanged Colubrids

Most rear-fanged colubrid snakes are harmless to man but in Africa two species have spectacularly demonstrated their lethal capacities by killing world famous herpetologists. A few other species have delivered worrying snakebites.

Boomslang *Dispholidus typus*

In 1957, American herpetologist Dr Karl P. Schmidt was bitten by a juvenile boomslang and died of a brain haemorrhage and respiratory failure. Rear-fanged snakes were not considered dangerous and there was no antivenom. The boomslang is a large-eyed, variably patterned, diurnal treesnake. Its venom, designed to kill chameleons and small birds, is also highly toxic to man. The short head and wide gape of the boomslang (p.15) enables it to bite easily. The defensive posture of the boomslang involves inflating its throat and flashing the contrastingly dark skin between its scales.

Range: Sub-Saharan Africa • Max. length: 1.5–1.8m • Venom: Haemotoxin, coagulant and haemorrhagin, causing bleeding and renal failure; several fatalities • Habitat: Savanna and thicket woodland • Prey: Chameleons, birds and frogs • Reproduction: Oviparous; 8–25 eggs • Similar species: Harmless treesnakes (*Thrasops* spp.).

Savanna twigsnake *Thelotornis capensis*

Generally inoffensive, the diurnal twigsnakes have caused deaths, including that of the German herpetologist Dr Robert Mertens in 1975. Four species are recognised. All are extremely slender treesnakes with elongate heads and cryptic patterning. They approach prey in a gentle swaying manner, the forked tongue protruding but motionless. The pupils are horizontal to enhance its ability to stalk camouflaged lizards. Defensive behaviour involves flattening the head into a lance-shape. There is no specific antivenom, boomslang antivenom does not work.

Range: East Africa from Somalia to Namibia and South Africa • Max. length: 0.6–1.0m • Venom: Haemotoxin, coagulant and haemorrhagin, causing bleeding and renal failure; several fatalities • Habitat: Savanna woodland. • Prey: Chameleons, frogs and birds • Reproduction: Oviparous; 8–10 eggs • Similar species: Three other twigsnake species (*Thelotornis* spp.).

Blanding's treesnake *Toxicodryas blandingi* (p.11)

Blanding's treesnake is an aggressive nocturnal snake that readily adopts a threat posture. Removed from the Asian genus *Boiga*, it may be more dangerous than any Asian catsnake. I received a bite from a large specimen that caused considerable muscular pain and difficulty in walking for several hours.

Range: West and central Africa • Max. length: 2.0–2.4m • Venom: Postsynaptic neurotoxin and anticholesterase; few snakebites, no fatalities • Habitat: Rainforest and woodland • Prey: Lizards, birds and their eggs, rodents and bats • Reproduction: Oviparous; 7–14 eggs.

Top: *A Boomslang (*Dispholidus typus*) killed the famous American herpetologist Dr Karl Schmidt in 1957. The large mouth and short head easily enable this rear-fanged treesnake to bring its fangs into play when it bites a human.*

Above: *A Savanna twigsnake (*Thelotornis capensis*) killed the eminent German herpetologist Dr Robert Mertens in 1975. Unfortunately the boomslang antivenom, produced in response to his friend Karl Schmidt's death, does not treat twigsnake bite.*

TROPICAL ASIA

Tropical Asia comprises South Asia (Pakistan, India, Sri Lanka, the Maldives, Nepal, Bhutan and Bangladesh), Southeast Asia (Myanmar [Burma], Thailand, the Indian-owned Andaman and Nicobar Islands, Vietnam, Laos, Cambodia, Malaysia, Singapore, Indonesia, Brunei and the Philippines) and Eastern Asian (southern China, Hainan Island, Hong Kong, Macao, Taiwan and the Japanese-owned Ryukyu Islands), a zoographical region known as the Oriental zone.

This heavily populated region is ecologically diverse, from Himalayan snow and rock to coral reef-fringed Pacific and Indian Ocean islands. There are smaller mountain ranges; the Ghats of India with their ancient rainforests, the peaks of Laos, Vietnam and Myanmar and towering Mt Kinabalu on the island of Borneo, arid habitats; the Deccan plain of India and Thar Desert in Pakistan, and great rivers; the Indus, Ganges, Brahmaputra, Irrawaddy and Mekong, with rocky gorges and wide grassy flood-plains.

Despite extensive logging, rainforests survive in India, the Andamans, Borneo, Sumatra and Malaysia, tropical deciduous and monsoon forests occur from Thailand to the Philippines and Java, bamboo forests are found in the hills and mangrove entanglements fringe coastlines, especially in Bhitarkanika, Orissa and the Sundabans, Bangladesh. Man-made monocultures include Myanmar's teak forests, India's tea plantations, and Sri Lanka's rice paddies. Of the twenty countries within Tropical Asia only the Maldives lacks terrestrial snakes. Tropical Asia has a population of 1.5 billion and over 100 species of venomous snakes. At the last estimate 25-35,000 Asians died of snakebite each year. Of the world's four highest annual death tolls, three: India, Sri Lanka and Myanmar, occur here.

Right: *At approximately 6.0m maximum length, the King cobra* (Ophiophgus hannah) *is easily the longest venomous snake in the world.*

Asian Cobras

The cobras are indelibly associated with Asia despite the fact they demonstrate greater diversity in Africa. The Afro-Asian cobra genus *Naja*, is a latinisation of the Sanskrit word *naia*, for 'snake'.

Common Cobras

Until recently all Asian cobras, except the king cobra, were considered subspecies of single species, *Naja naja*, but they are now recognized as eleven separate species. The common cobra complex includes spitting cobras and non-spitting cobras, but the boundaries between the two are not as easily defined as with African cobras. Whether a cobra can spit venom in defense is a product of its biology and morphology. Their fangs and venoms are different. Non-spitting cobras possess primarily neurotoxic venom whilst spitter venom tends to be more cytotoxic venom, to cause maximum effect, pain and blindness when it contacts the cornea of the eye. The division between spitters and non-spitters may be geographical. Middle Eastern and South Asian cobras are non-spitters but cobras from the Indo-Malay-Philippine archipelago, which has only one common cobra per island, are spitters. The exceptions are the non-spitting Andaman cobra and males of the northern Philippine cobra. The situation on the Southeast Asian mainland is more confused with spitters and non-spitters occurring in the same locations. Some populations of spitters are reluctant to spit and it is easy to be caught unawares by a Southeast Asian cobra that has never spat before.

Because common cobras prey on rodents, snakes and amphibians they are commonly encountered around human habitations and in agricultural areas, drawn by the presence of prey and diversity of habitats. Despite the high numbers levels of snakebite morbidity and mortality they cause throughout Asia, cobras are highly respected in both Buddhist and Hindu cultures. Buddhists see the cobra as the protector of Buddha, sheltering him from heavy rain and receiving his touch in return. In Sri Lankan cobras the spectacle marking is seen as Buddha's thumb and finger print

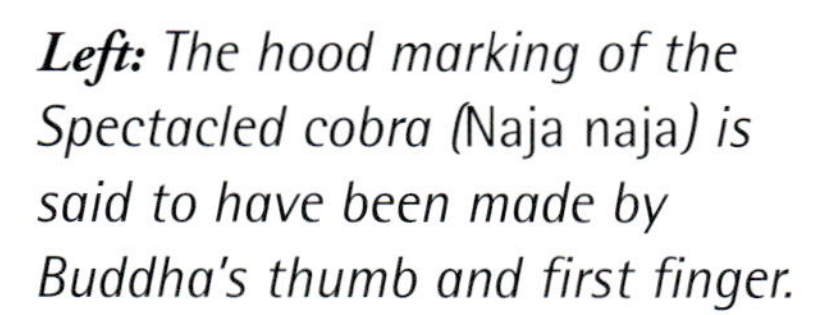

Left: *The hood marking of the Spectacled cobra (*Naja naja*) is said to have been made by Buddha's thumb and first finger.*

Right: *Legend has it that the Monoculate cobra's (*Naja kaouthia*) hood marking is made by the thumb-print of Buddha.*

while in Thailand, Buddha's thumb-print is represented by the single monocle marking of the Thai cobra. Many Buddhists will not kill cobras, allowing them to remain in the paddi-fields and rice stores, eating rats and sometimes biting people. Buddha was also reputed to use the cobra as his vehicle for exacting justice to wrongdoers.

Snake charmers used to be a common sight in many Asian towns and cities. There are good and bad snake charmers, those who care for their snakes and those who abuse them by pulling fangs or sewing up their mouths. Some of the most experienced snakemen are the Irula tribe of southeastern India who used to make their livings by catching snakes and selling them for skins but who now catch venomous snakes for snakebite research programs, supplementing their livelihoods with rat removal services. Snake charming is a dying trade in India due to regulations concerning cobras in public places. Today charmers may be arrested and their snakes confiscated.

Mongooses are a natural predator of cobras but they are not the main threat to their survival. In parts of Asia, especially China and places with large populations of ethnic Chinese, cobras are sought as food and for the production of traditional remedies for increased human virility or longevity. In some cities a visit to a snake restaurant is a must-do for tourists. These massive harvests of cobras and other snakes also do immense harm to the balance of nature by removing the natural predators of the rice-stealing, disease-carrying rats.

Indian or Spectacled cobra *Naja naja*

The Indian cobra, a non-spitter, is also known as the spectacled cobra because of the mark on the rear of its hood that resembles a pair of wire-rim spectacles. Southern specimens are brown, Sri Lankan cobras may be speckled or banded with cream, while individuals from Pakistan and Nepal are jet black with their hood markings almost obscured. Being dark is an advantage for snakes living in cooler northern climes such as the Himalayan foothills. Indian cobras hunt rats in paddy-fields and around buildings, especially at night, and come into contact with humans. There are thousands of snakebite accidents annually. Without antivenom they may rapidly lead to respiratory paralysis and death.

Range: Nepal, Pakistan, India and Sri Lanka • Max. length: 1.2–2.0m • Venom: Postsynaptic neurotoxins and localized cytotoxic necrosis (tissue-death); numerous human fatalities • Habitat: Monsoon forest to paddy-fields • Prey: Rodents, toads and snakes • Reproduction: Oviparous; 8–32 eggs.

Thai or Monoculate cobra *Naja kaouthia*

The Thai cobra, a non-spitter, is also known as the monoculate cobra because its hood marking resembles a round monocle. The common Southeast Asian cobra, it also occurs along the Ganges valley into northern India and Nepal. Like the Indian cobra, this common inhabitant of cultivated fields is a source of many human snakebite accidents and deaths. In Buddhist Myanmar and Thailand the round mark on the hood of this cobra is said to represent Buddha's thumb-print.

Range: Mainland Southeast Asia, northern India and southern Nepal • Max. length: 1.4–2.3m • Venom: Postsynaptic neurotoxins and localized cytotoxic necrosis (tissure-death); numerous human fatalities • Habitat: Tropical forest to paddy-fields. • Prey: Rodents, toads and snakes • Reproduction: Oviparous; 8–45 eggs.

"I have caught many cobras in Indian paddy fields and outhouses, woodland and coconut plantations in Sri Lanka and holes in riverbanks in Nepal. They all responded boldly to being disturbed, raised a hood, hissing and lunging. This classic response is the prime ingredient in one the most abiding images or rural India, that of the snake charmer with his basket of cobras playing a gourd flute to charm his deaf charges into doing what comes naturally when faced with a visual threat and a gullible audience. Sri Lanka, India and Nepal.

Indochinese spitting cobra *Naja siamensis*

The distinctive Indochinese spitting cobra was, for many years, misidentified as various other species despite being a snake of considerable medical importance. Venom spat into the eyes caused numerous accidents and snakebites from this species resulted in the loss of large areas of tissue and terrible scarring. It is a strange fact that this species was a common and familiar species in captivity before it was recognized and described as a valid species by scientists.

Range: Thailand, Cambodia and southern Vietnam • Max. length: 1.0–1.2m • Venom: Postsynaptic neurotoxins and localized cytotoxic necrosis (tissue-death); few human fatalities • Habitat: Lowlands and cultivated hill country • Prey: Rodents, toads and snakes • Reproduction: Oviparous; 13–19 eggs.

Above: *The Indochinese spitting cobra (*Naja siamensis*) was, until recently, confused with other cobra species.*

Right: *This Southern Indonesian spitting cobra (*Naja sputatrix*) was captured on Komodo, home to three species of venomous snakes as well as the famous dragons.*

Southern Indonesian spitting cobra *Naja sputatrix*

The southern Indonesia spitter is a unicolour black, yellow or brown cobra found in the dry forests of Java and the Lesser Sunda islands. Cobras have also been reported from the island of Sulawesi and they may prove to be this species. On Komodo and Flores the spitting cobra faces a dangerous foe in the shape of the Komodo dragon that would make short work of even an adult spitter. It has been reported that Komodo specimens are reluctant to spit but my experiences in the field suggests differently.

Range: Java to Komodo and Flores, possibly Sulawesi • Max. length: 1.0–1.5m • Venom: Postsynaptic neurotoxins and localized cytotoxic necrosis (tissue-death); few human fatalities • Habitat: Lowlands and cultivated hill country • Prey: Rodents, toads and snakes • Reproduction: Oviparous; 13–19 eggs.

"I caught a small spitting cobra near a Komodo dragon nest and I found it extremely eager to demonstrate its prowess while I photographed it, hitting me from the hairline to the beard. Fortunately I was wearing eye protectors. Rintja, Komodo Islands, Indonesia.

"I was trying to photograph a 'Chinese cobra', not known to be from a spitting population, when it spat at me. I turned my head to the left but received both barrels in my right eye. Although very painful I was able to bathe my eye repeatedly and thoroughly with water for almost one hour and managed to remove most of the venom. My first aid was successful. Later examination at an eye hospital showed that although the cornea was pockmarked it was not severely damaged and it recovered in a little over one week.

Other common cobras: Caspian cobra (*N. oxiana*) from Iran to Pakistan; Andaman cobra (*N. sagittifera*) from Andaman Islands, Bay of Bengal; Chinese cobra (*N. atra*) from Vietnam, China and Taiwan; Equitorial spitting cobra (*N. sumatrana*) from Sumatra to Palawan; North Philippine cobra (*N. philippinensis*) from Luzon and Mindoro; Southern Philippine cobra (*N. samarensis*) from Samar and Mindanao; and as recently as 2001 the Mandalay spitting cobra (*N. mandalayensis*) was described from Myamnar (Burma).

King Cobras

If there is one species of snake that is respected above all others, not just for its potential danger and size, but also for its apparent 'intelligence' it must be the king cobra.

King cobra *Ophiophagus hannah*

The king cobra is the longest venomous snake in the world. Although only a single widespread species is recognized for all king cobras, whether from Thailand, southern China, the Philippines, Indonesia or India, an Indian herpetologist is nearing the completion of a revision of the species throughout its entire range that will undoubtedly lead to it becoming several separate geographical species.

The king cobra is unique amongst snakes in that it not only builds a nest of leaves to incubate its eggs, it also remains with its eggs to guard them, even against elephants which may be killed by trunk-tip bites. Capable of raising one-third of its length off the ground, when it rears up and hoods, and injecting a large quantity of venom when it bites, an adult king cobra is not only the most impressive of all snakes but also one of the most dangerous. Coupled with its reputation for intelligence, growling when annoyed and showing its enemy a 'bared fang' to emphasize its intentions, the king is a very special snake indeed. It can be distinguished from all other snakes by a pair of extra scales on the back of its head called occipital scales.

Although widespread throughout Southeast Asia, its South Asian distribution is patchy. The king cobra is found in northeastern India, in Assam and Bengal, and Bangladesh, from where it ranges both westwards along the Ganges valley and the terai-wooded foothills of Himalayan Nepal and northern India as far as the Pakistan border, and south from Bengal to the mangrove swamps of Orissa. In Orissa, kings are reputed to nest in trees, which was a sight I hoped but failed to witness when I visited the king cobra 'hot-spot' of the Bhitarkanika swamps. There are also isolated populations in the endangered montane rainforests of the Western Ghats, southern India. Here kings are relatively common and extremely variable in coloration, from pale brown to black with cream chevrons on the back of the long narrow hood.

The king cobra's habit of feeding on other snakes, especially the large dharman ratsnakes (*Ptyas mucosus*), is reflected in its scientific name *Ophiophagus* (*ophio* = snake; *phagus* = eater). The king cobra is responsible for relatively few snakebites, compared to common cobras, kraits and vipers. Some bites are to dancers who handle the snakes as part of their performances, especially in Myanmar. Although king cobra venom is not as toxic as that of a common cobra, a large king can inject a huge amount with rapidly fatal results. Average venom yield is 200–500mg while the estimated lethal human dose is 12mg. The lack of rural snakebites may be due to the secretive nature of king cobras. They tend to

inhabit less-populated forested regions where human interactions are less frequent, whereas the common cobras feed on rodents and amphibians and so more frequently come into conflict with humans on agricultural land or around homes.

Range: India to southern China, south to Sulawesi (Indonesia) and the Philippines • Max. length: 4.0–6.0m • Venom: Postsynaptic neurotoxin • Habitat: Dry and tropical forests, plantations, mangrove swamps, grassland and bamboo thickets • Prey: Mostly snakes, including ratsnakes, pythons and cobras, occasionally lizards and small mammals • Reproduction: Oviparous; 20–51 eggs, females guard their nests.

"Local beliefs about king cobras are many. It is believed if 100 people see a king cobra it will die and also that they come to villages to die. The first king I encountered in India was a 3.5m male on the outskirts of a village. It was dying. I subsequently captured a 3m female king behind a temple. It was in the process of catching, killing and eating a large dharman ratsnake. When testing the senses of this specimen later I turned to find her hooding and gazing directly into my eyes, and when she caught my attention I saw her pupils focus on my face and she started to sway. I started to sway with her and was sure she was asking me why I had caught her. I felt extremely guilty and determined to release her as soon as possible, which I did, in a remote forest creek. It was the most spiritual moment of my life, I am sure she was trying to communicate with me, something I have not experienced with any other snake, nor expect to do so again.

Western Ghats, Karnataka, southern India.

Opposite: *The King cobra (*Ophiophagus hannah*) is characterised by its enlarged occipital scales, visible on the rear of this cobra's head, and often has an inverted chevron pattern on the back of its neck.*

Above: *Juvenile King cobras (*Ophiophagus hannah*) may be brightly marked but these colours often fade as they mature. Baby king cobras are very alert and will hood at the slightest provocation.*

Other Asian Elapids

Although cobras may dominate the elapid snake fauna of Asia, 28 species of non-cobra elapids are also found in the region. Although the oriental coralsnakes are not a significant snakebite threat, the same cannot be said of the kraits which may account for as many fatalities as the cobras.

Kraits

The kraits are an entirely Asiatic group, unlike the Asian cobras they have no close African relatives. Kraits are also the 'Jekyll and Hydes' of the snake world, shy and retiring during the day, deadly at night. Most kraits also have nonvenomous mimics that will also bite if handled. Confusion over the identity of the snake concerned, or even whether a bite has taken place, combined with a lack of pain or swelling which give no clues to the seriousness of the bite, may lead to fatal delays in seeking medical attention. Out of 12 species, approximately half are responsible for thousands of human fatalities annually, while the bites from the remaining species are un-known. Antivenom is not available for every species.

Common krait *Bungarus caeruleus*

Common kraits hunt nocturnally and may enter human dwellings, burrowing under people sleeping on the floor. At night they are quick to bite, often without waking their victims since the venom is painless. Victims wake later, paralysed, or die in their sleep. The cosmopolitan common krait, a shiny brown or blue-black snake, with or without cream bands, is one of the most dangerous of all Asian snakes. In some parts of Pakistan the common krait, and its relatives kill more people than cobras or vipers and it is also responsible for a huge death toll in Sri Lanka. Venom yield is 30mg, lethal human dose is 2.5mg. The simple expedient of sleeping off the ground would greatly reduce krait bites.

Range: Pakistan, Nepal, India and Sri Lanka • Max. length: 1.2–1.7m • Venom: Pre- and postsynaptic neurotoxins
- **Habitat: All lowland habitats from gardens to rainforest**
- **Prey: Primarily snakes, also amphibians and small mammals**
- **Reproduction: Oviparous; 6–15 eggs.**

Banded krait *Bungarus fasciatus*

The widespread banded krait is easily recognized because it is one of the few snakes with a triangular body, in cross-section, and a distinctive raised vertebral ridge further emphasized by an enlarged row of vertebral scales. Boldly patterned with broad yellow and black or brown bands, with a flattened head, small eyes and a blunt-rounded tail tip, it is difficult to confuse this snake with any other species. It is also responsible for snakebites though fewer deaths than the common krait. Venom yield is 43mg, lethal human dose is 10mg.

Range: Northeastern India to China and Indonesia • Max. length: 1.8–2.1m • Venom: Pre- and postsynaptic neurotoxins. • Habitat: Many low-lying habitats from open to forested country, especially near water • Prey: Mostly snakes and their eggs, also amphibians and small mammals • Reproduction: Oviparous; 3–14 eggs.

"Kraits are amongst the most frustrating snake to photograph. Any illumination will cause them to hide their heads under their coils and their habit of rapidly striking to the side can be very disarming when working close with a macro-lens. Patience is essential if one is to avoid getting bitten. **Royal Bardia National Park, western Nepal.**

Other dangerous kraits: The common krait is related to the Sind krait (*B. sindanus*) of Pakistan. Dangerous species in Southeast Asia include the many-banded krait (*B. multicinctus*) and the Malayan krait (*B. candidus*). The strangest krait is the brightly coloured yellow-headed krait (*B. flaviceps*) of Borneo, Sumatra and Java.

Left: *The Common krait (*Bungarus caeruleus*): shy and retiring during the day, lethal at night.*

Opposite: *The Blue long-glanded coralsnake (*Calliophis bivirgata*) has venom glands that extend along the anterior portion of the body.*

Oriental Coralsnakes

There are 14 species of oriental coralsnakes which may be split into the long-glanded coralsnakes, slender coralsnakes, and Chinese-Philippine coralsnakes. Most are small, secretive and inoffensive but a few have caused human fatalities.

Blue long-glanded coralsnake *Calliophis bivirgata*

The venom glands of long-glanded coralsnakes extend down the first third of the body. The largest and most dangerous is the Blue long-glanded coralsnake which is overall blue-black with a pair of light blue stripes down each side, a bright red head, tail and underbelly and a faded red vertebral stripe. Defensive posture consists of coiling the tail into a red helix. A rare rainforest species, it has highly toxic venom. A Singapore man died five minutes after being bitten on the toe in the shower.

Range: Southeast Asia from Myanmar to Java • Max. length: 1.8m • Venom: Possibly neurotoxic but little known; at least two human fatalities • Habitat: Lowland rainforest • Prey: Snakes, including other coralsnakes, legless lizards and frogs • Reproduction: Oviparous; 1–3 eggs • Similar species: Striped long-glanded coralsnake (*C. intestinalis*) - Thailand to Philippines, six species of slender coralsnakes (*Calliophis* spp.) and the yellow-headed krait (*Bungarus flaviceps*).

MacClelland's coralsnake *Sinomicrurus macclellandi*

MacClelland's coralsnake, a dark red snake with black rings and a white and black head, is the most widespread Chinese coralsnake and the only one to have caused a fatality. A German herpetologist died eight hours after neglecting a coralsnake bite in Nepal, demonstrating the danger posed by small elapids and snakebites with an absence of pain or swelling.

Range: Nepal to Southeast Asia, China and Taiwan • Max. length: 0.8m • Venom: Possibly neurotoxin and/or myotoxin but little known, at least one human fatality • Habitat: Hill jungles • Prey: Snakes and legless lizards • Reproduction: Oviparous, 4–14 eggs • Similar species: Four other Chinese coralsnakes (*Sinomicurus* spp.), from Taiwan and Ryukyu Islands, and Philippine coralsnake (*Hemibungarus calligaster*).

Asian True Vipers

True vipers not well represented in Tropical Asia. The Russell's viper is the one truly Asian species. The Pakistan and Indian carpet vipers are Asian representatives of a wide-ranging Middle Eastern-North African complex.

Pakistani carpet viper *Echis sockureki*

Carpet vipers are brown with a patterning of darker blotches. The head is short and the eyes large. The unusual arrangement of the keeled, serrated scale rows along the flanks are characteristic of these vipers. When threatened the viper forms its body into a series of concentric curves and moves backwards, rubbing one series of serrated, keeled scales against another to generate a sawing sound. The sound created had led to the alternative name of saw-scale viper. As a defence, the sawing sound is as effective as the rattle of a rattlesnake, and (as with the rattlesnake) this method also conserves moisture that could be lost by hissing.

Carpet vipers are among the most dangerous snakes in the arid North African and west Asian region. In Asia they vary greatly in size and the snakebite risk they pose. The Pakistan carpet viper may be the most dangerous snake within its range, killing hundreds of people every year. It is extremely common, often thousands may be found in a single locality. The strike consists of a short jabbing bite and death may occur within days through cerebral haemorrhage, gastrointestinal haemorrhage or renal failure. By contrast the diminutive Sri Lankan carpet viper (0.3m) does not feature in that country's otherwise high snakebite statistics. Similar species: Indian and Sri Lankan carpet vipers (*E. carinatus* ssp.) and eight Afro-Middle Eastern species.

Range: Pakistan and northeastern India • Max. length: 0.8m. Venom: Haemorrhagin and procoagulant; many human fatalities • Habitat: Semi-desert to scrub • Prey: Lizards, toads, nestling birds and locusts • Reproduction: Viviparous; 8-28 neonates (sometimes oviparous).

Russell's viper *Daboia russelii*

Named in honour of Patrick Russell, an 18th century Scottish physician and the naturalist to the East India Company, who conducted the first detailed studies of Indian snakes, this is a very dangerous nocturnal snake, responsible for thousands of snakebite fatalities throughout its fragmented range. The main

Below: *The Pakistani carpet viper (*Echis sockureki*) may be locally very common, and very deadly.*

Opposite: *The Russells viper (*Daboia russelii*) possesses a complicated and highly toxic venom.*

populations occur in mainland Asia but isolated populations exist in Taiwan and Indonesia. Although the snakes look similar they exhibit considerable variation in their venom composition, which has a serious effect on snakebite treatment. Current scientific opinion suggests that the Indochinese Russell's viper (*D. r. siamensis*) may warrant specific status.

The Indochinese race are orange in colour, with a series of large, bold, black edged red lozenge-markings down the centre of the back and a second row down either side. They can be distinguished from the Indian race by the presence of smaller dark spots between the lozenges. An extremely irascible snake, it hisses loudly and is quick to bite with a rapid and far-reaching strike.

Particular snakebite problems exist with this snake in Myanmar (Burma), Indonesia, Taiwan and Sri Lanka because the antivenom available to treat snakebites is not based on the venoms of the native Russell's vipers and therefore may not treat all the life-threatening symptoms from its bite. Sri Lanka serves as a good example. When a large area in the north underwent a massive irrigation, rice-planting and trans-migration program 30 years ago the snakebite statistics escalated. Only the imported Indian antivenom was available but Sri Lankan viper venom can cause cerebral haemorrhages and renal failure, problems not fully addressed by the Indian antivenom unless large quantities are administered. Today the Russell's viper, along with the cobra and the kraits, is a major cause of snakebite in Sri Lanka, killing hundreds of villagers every year. A programme has been initiated to produce a specific Sri Lankan antivenom.

Range: Pakistan to Sri Lanka, Myamnar to Cambodia and China, Taiwan, Indonesia (Java to Flores) • Max. length: 1.2m • Venom: Procoagulant, possibly anticoagulant, haemorrhagin, presynaptic neurotoxin in South Asia; myotoxin in Sri Lanka. Yield: 130–250mg/kg, fatal human dose 40–70mg. Numerous human fatalities • Habitat: Most habitats, especially paddy-fields, not rainforest • Prey: Rodents • Reproduction: Viviparous; 20–40 neonates.

"In Sri Lanka the Russell's viper is known as *tith polonga* but to the northwest lives a snake locals call the *getapolonga*. They claim it resembles a young *tith polonga*, but its bite kills in minutes. While searching for specimens I examined a dead *getapolonga* but it did not appear any different from a juvenile Russell's viper to me. Is there a local population of Russell's viper even more deadly than the typical Sri Lankan race? Anuradhapura, Sri Lanka.

"I accompanied the Thai forestry police in a raid on an illegal reptile skin and meat factory. In one room I found a dozen very noisy sacks and when I opened them I found over 100 live Thai Russell's vipers destined for harvest, probably for their gall bladders, which are believed to have beneficial quasi-medical qualities. Bangkok, Thailand.

Fea's Viper

The subfamily Azemiopinae contains one living species, Fea's viper from the montane region of north Myamnar and southern China. Although it lacks pits it is thought to represent a closely related sister-taxa of the widespread Asian-American pitvipers.

Fea's viper *Azemiops feae*

One of the most fascinating snakes in Asia, Fea's viper, was originally described as an elapid. It has smooth scales and the primitive 'nine-plate arrangement' of head scales, more usually associated with colubrids or elapids. A mix-match of characters. it also possesses typically viper-like, vertical elliptical pupils. Fea's viper's colour patterning, dark grey with a widely-spaced series of black-edged, orange bars or rings, and a pale orange, almost white, head with darker orange stripes running back onto the nape, must make it one of the strangest vipers alive. The natural history of Fea's viper is poorly documented apart from occasional field observations. It is also difficult, but not impossible, to maintain in captivity.

Range: Northern Myanmar and Vietnam, southern Tibet and southwestern China • Max. length: 0.7–1.0m • Venom: Believed to be neurotoxin and anticoagulant; mild human bites recorded • Habitat: Karst and cloud forest and mixed bamboo and fern forest from 800–2000m • Prey: Small mammals, especially grey shrews • Reproduction: Probably oviparous.

Below: *The curious and little known Fea's viper (*Azemiops feae*) from Myamnar, is believed to be the close relation of the ancestral pitvipers.*

Opposite: *The Malayan pitviper (*Calloselasma rhodostoma*) causes many snakebites in Vietnamese and Thai forests but barely enters Malaysia, so it is rather misnamed.*

Asian Pitvipers

Pitvipers evolved in Asia and spread across the Bering Strait into the Americas, so it comes as no surprise that there is a huge diversity of pitvipers in tropical Asia where they have evolved to occupy most terrestrial habitats except deserts. The region contains over 50 pitvipers, both stout-bodied, cryptically-patterned terrestrial species, and slender, arboreal species, and oviparous as well as viviparous species.

Terrestrial Asian Pitvipers

The forest-floor pitvipers include those with the primitive colubrid nine-enlarged-plate arrangement, such as the Malayan, Hundred-pace, Humpnose and Himalayan pitvipers, and the more advanced, fragmented, granular scale arrangement of the Mountain pitvipers.

Malayan pitviper *Calloselasma rhodostoma*

Greatly feared, the mis-named Malayan pitviper only enters extreme northwestern Peninsular Malaysia and is really a Thai-Vietnamese species. As an Asian pitviper it is unusual – its scales are smooth, and it lays eggs. Snakebite fatalities occur from cerebral haemorrhage or blood poisoning due to infection of necrotic wounds. Its sedentary nature and cryptic patterning greatly increase the risk posed. A mainland dry forest species, absent from rainforests, it also occurs in arid Java, a distribution pattern reflecting that of the Russell's viper which also persisted in the small Javanese enclave. Malaysian pitvipers aestivate through the annual dry seasons, becoming active at the start of the monsoon. This pitviper has also saved lives because a drug used to prevent blood-clotting is produced from its venom.

Range: Thailand, Vietnam and Cambodia northern Malaysia, also Java • Max. length: 1.0m • Venom: Procoagulant, haemorrhagin and cytotoxin; many human fatalities • Habitat: Dry lowland and hill habitats, including coffee and rubber plantations • Prey: Small mammals, birds, lizards and amphibians • Reproduction: Oviparous; 10–35 eggs • Similar species: Hundred-pace pitviper (*Deinagkistrodon acutus*).

Hundred-pace pitviper *Deinagkistrodon acutus*

Bearing a name that is said to reflect how far victims can walk before succumbing to the bite, the hundred-pace pitviper is also known as the Chinese copperhead due to its superficial resemblance to the American copperhead. Other local names translate as 'snake of 100 designs', a reference to its cryptic patterning, 'sharp-nosed pitviper', because the tip of its snout is strongly upturned, and the worrying local Chinese name of 'five-pacer'. The hundred-pace pitviper is the Chinese equivalent of the Malayan pitviper, albeit with mostly keeled scales. Its main predator may be humans because it is harvested for food and traditional medicines.

Range: Northern Vietnam, China and Taiwan • Max. length: 1.3–1.6m • Venom: Procoagulant, haemorrhagin and cytotoxin; human fatalities recorded • Habitat: Dry rocky and wooded hills • Prey: Small mammals, birds, lizards and amphibians • Reproduction: Oviparous; 6–35 eggs • Similar species: Malayan pitviper (*Calloselasma rhodostoma*).

Indian humpnose pitviper *Hypnale hypnale*

The Western Ghats of southwestern India are home to many unusual species, including several pitvipers. The small cryptically coloured, brown, leaf-litter dwelling Indian humpnosed pitviper, with its slightly upturned snout, is confined to this region and the island of Sri Lanka. This species is a common cause of snakebites among people gathering firewood or clearing brush around fields and gardens but there have been only a few fatalities.

Range: Southwestern Indian and Sri Lanka • Max. length: 0.5–0.6m • Venom: Procoagulant causing irreversible renal failure; human fatalities in Sri Lanka • Habitat: Forests and plantations • Prey: Small mammals and lizards • Reproduction: Viviparous; 4–10 neonates • Similar species: Sri Lankan humpnose pitviper (*H. nepa*), Wall's humpnose pitviper (*H. walli*) from central Sri Lanka.

“Humpnose pitvipers can be extremely common. I found six in ten minutes in a small patch of undergrowth only a few square meters in size. Although a fairly sedentary species I also found a specimen crossing a dirt road at night in southern India. Nikaweratiya, Sri Lanka and W. Ghats, India.

Below *The Mountain pitviper (*Ovophis monticola*) inhabits mountain ranges between 1000-2000m altitude.*

Opposite top *The Hundred-pace pitviper (*Deinagkistrodon acutus*) so named, legend tells us, because that's as far as victims get before dying.*

Opposite bottom: *The little Indian humpnose pitviper (*Hypnale hypnale*) can be very common in gardens, I found six in ten minutes in a few square metres of brush.*

Himalayan pitviper *Gloydius himalayanus*

Several pitvipers occur in Nepal, both in the terai forests along the Indian border and up into the Himalayas but the Himalayan pitviper is the only member of the Eurasia genus *Gloydius* to enter Tropical Asia. Whether its distribution can be called 'tropical' is debatable since it has not been recorded from below 1500m and holds the record as the world's highest distributed snake, with an altitudinal record of 4876m, from the foot of the Dharmsalah Glacier in the western Himalayas. The pitviper may hibernate for up to eight months at extreme altitudes. Himalayan pitvipers are grey or brown with darker blotches to help them blend in among the pine needle leaf litter.

Range: Himalayas from north Pakistan and India to Nepal • Max. length: 0.5–0.9m • Venom: Nothing known other than localized effects; no human fatalities • Habitat: Grassy, rocky and coniferous hillsides • Prey: Mice, lizards, frogs and centipedes • Reproduction: Viviparous; 3–7 neonates • Similar species: Siberian pitviper (*G. halys*) and mamushis (*Gloydius* spp.) of Japan, China and Korea.

Mountain pitviper *Ovophis monticola*

A stocky terrestrial pitviper, the mountain pitviper is brown with squarish darker brown markings and a dark brown centre to its head. Being a snake confined to mountainous regions between 1000 and 2000m, the widespread distribution of the species and the large expanses of low-lying country between the various populations, has resulted in several subspecies being recognized. Although less frequently encountered by humans than lowland pitvipers, the mountain pitviper is capable of delivering worrying snakebites although few fatalities are recorded.

Range: Nepal to Sumatra, China and Taiwan • Max. length: 0.5–1.1m • Venom: Presumed haemorrhagin and procoagulant but little known; one human fatality recorded • Habitat: Rocky, oak and coniferous hillsides • Prey: Mice, birds, lizards and frogs • Reproduction: Oviparous; 4–18 eggs • Similar species: Okinawa hime-habu (*O. okinavensis*) from the Ryukyu Islands, Japan, and Chasen's pitviper (*Garthia chaseni*) from Mt Kinabalu, Borneo.

Okinawa habu *Protobothrops flavoviridis*

The name *habu* is usually reserved for the species from China, Taiwan and the Ryukyu Islands. The Okinawa habu, the longest Asian viper, was once responsible for a serious snakebite problem in the Ryukyu Islands. Habus entered homes and outbuildings in search of prey and snakebites occurred to almost anybody, but those at most risk were sugar-cane workers. Thriving on a diet of rats and living alongside humans, the pitviper population grew rapidly until it was one of the major causes of serious and fatal snakebites in the world. An entire range of snake barriers, traps and deterrents were introduced to halt the snakes, as well as a programme of removal and eradication of both the rodent prey and the pitvipers, the latter using biological controls, from parasites to snake-sniffing dogs and the Indian mongoose, and chemical agents acting through the skin. The result has been a drop in the incidence of snakebites and the development of a specific antivenom has also greatly reduced the fatality rate following bites. The Okinawa habu has become one the most studied venomous snakes in the world and probably the one with the largest database of snakebite statistics, documented religiously since 1970 in the Japanese Snake Institute journal *The Snake*.

Range: Amami and Okinawa island groups, Ryukyu Islands, Japan • Max. length: 2.0–3.0m • Venom: Cytotoxin and haemorrhagin, but shock was the cause of high death rate before antivenom • Habitat: All habitats from forests to sugar-cane fields • Prey: Rodents • Reproduction: Oviparous; 5–17 eggs.

Jerdon's lancehead *Protobothrops jerdoni*

The semi-arboreal Asian lanceheads and habus strongly resemble the lancehead pitvipers of Latin America. The boldly patterned Jerdon's pitviper is a lime to olive green snake, every scale edged with black, with a series of black-edged brown saddles over the back, and black stripes on its broad green head. Three subspecies of this semi-arboreal, crepuscular (active at twilight) pitviper, which occurs to an altitude of 2800m, are recognized.

Range: Nepal to Vietnam and northern China. • Max. length: 1.3–1.4m • Venom: Procoagulant, possibly haemorrhagin, little known; no human fatalities recorded • Habitat: Montane forests and scrubland. • Prey: Small mammals and birds • Reproduction: Viviparous; 2–8 neonates • Similar species: Two more Asian lanceheads and three other habus (*Protobothrops* spp.) from India to the Ryukyu Islands.

Above: *The Okinawa habu (*Protobothrops flavoviridis*) used to cause many snakebites but a program of eradication of both the habu and its rodent prey from around buildings has greatly reduced snakebite incidence and deaths.*

Asian and Bamboo Pitvipers

At least 36 species of arboreal and semi-arboreal Asian pitvipers were, until recently, included in the widespread genus *Trimeresurus* but they are now split between *Trimeresurus* and six new genera (*Cryptelytrops, Himalayophis, Parias, Peltopelor, Popeia* and *Viridovipera*). In many respects they mirror the palm-pitvipers and forest-pitvipers of tropical America and the bushvipers of Africa. Although most are arboreal, slender bodied with prehensile tails, including the many confusing green species often termed generically bamboo pitvipers, some species may also be found on the ground, especially the brown or grey-brown species. Accurate identification can be very difficult, especially when more than several species of bamboo pitviper occur in the same location. The only characters separating two or three sympatric species are finite differences in head scalation or the appearance of the male hemipenes (copulatory organs) that requires rather intimate examination of the snake. Most species are not highly dangerous but a few are of medical importance, having caused serious, even fatal snakebites.

White-lipped pitviper *Cryptelytrops albolabris*

The white-lipped pitviper is a typical bamboo pitviper. An attractive and widely available species, it is commonly kept in captivity but is often confused with other bamboo pitvipers, specially Pope's and the Chinese bamboo pitvipers. This is probably the bamboo pitviper most frequently implicated in snakebites, in the wild and captivity, with results ranging from mild envenoming to death. It is now believed this single species may actually be a species complex containing several, un-described species. The Sunda Island pitviper I was bitten by on Komodo (see page 109) was formerly considered a subspecies.

Range: Assam, India, to China, Thailand, Vietnam, Sumatra and Java • Max. length: 0.7–1.0m • Venom: Procoagulant; many bites with few fatalities • Habitat: Open tropical forest to bamboo thickets, cultivation and even urban gardens • Prey: Frogs, lizards, small birds and rodents • Reproduction: Viviparous; 4–25 neonates • Similar species: Pope's pitviper (*Popeia popeiorum*), Chinese pitviper (*Viridovipera stejnegeri*), green pitiper (*T. gramineus*), Sunda Island pitviper (*Cryptelytrops insularis*).

Malabar pitviper *Trimeresurus malabaricus*

The Malabar pitviper is only found in the ancient rainforests of the Malabar Coast in southwestern India where we were

Below: *The White-lipped pitviper (*Cryptelytrops albolabris*) is one of several confusing green bamboo pitvipers in South-East Asia.*

fortunate enough to find a specimen while searching for king cobra nests. Green to olive brown with a series of darker irregular blotches over its back, the Malabar pitviper is a variable and little known species.

Range: Western Ghats, southwestern India • Max. length: 1.0–1.1m • Venom: Only localized effects reported; no human fatalities • Habitat: Rainforest and dry forest • Prey: Tree frogs, geckos and shrews • Reproduction: Viviparous; litter size unknown.

Sri Lankan green pitviper *Trimeresurus trigonocephalus*

This highly arboreal pitviper, the only truly arboreal front-fanged venomous snake in Sri Lanka, is one of the most attractive species on the island, a kaleidoscope of green, cyan, yellow and black. Although widespread it is so inoffensive as to be considered virtually harmless by Sri Lankans.

Range: Sri Lanka • Max. length: 0.8m • Venom: Only localized effects reported with no human fatalities • Habitat: Forests, bamboo thickets and plantations • Prey: Frogs and lizards • Reproduction: Viviparous; 5–26 neonates.

"I was bitten by a small Sunda Island pitviper, (*Cryptelytrops insularis*) while I was filming in the Komodo Islands. I found the small turquoise specimen and took it back to the photographic set I had established on our boat, moored in the bay. I was bitten in the thumb through the snakebag. What ensued was a painful five hours while the swelling increased and we pondered whether to

arrange a medi-vac (medical evacuation). Although this species had already caused at least one human fatality, I considered that this juvenile specimen was too small to be a serious threat and since only one fang had penetrated, I decided to monitor the bite rather than dashing to hospital in Bali. By the next day I had recovered almost completely. Komodo Islands, Indonesia.

"While visiting the remote Batanes or Typhoon Islands in search of flying lizards I hoped to also find the endemic McGregor's pitviper (*Parias flavomaculatus mcgregori*), which reputedly occurs in two colour phases, yellow and white. After a long search in low montane forests at night I came across a juvenile specimen, almost as white as snow, crawling over the leaf litter. My guide told me the white ones are always on the ground and the yellow ones are in the trees. Batan, Batanes Islands, Philippines.

Opposite: *The Malabar pitviper (*Trimeresurus malabaricus*) is confined to the coastal forests of the West Ghats, southwest India.*

Below: *The Sri Lankan green pitviper (*Trimeresurus trigonocephalus*) is inoffensive and possesses a weak venom; it does not feature in snakebite statistics.*

Wagler's temple pitviper *Tropidolaemus wagleri*

Wagler's temple pitviper is one of most beautiful and variable of venomous Asian snakes. Coloration and patterning vary from juvenile to adult. Juveniles are green with scattered white and red-brown spots, and a brown stripe through the eye. Adult females are black, merging to green on the flanks, with a variable pattern of green and yellow spots, bars and bands. Adults also have especially broad, angular heads with the eyes high on the sides, and the snout slightly pointed. There is usually a broad dark stripe through the eye from tip of snout to the angle of the jaw.

Stout-bodied females can grow to over a 1m, while males are considerably smaller and slimmer. This pitviper is a typical sit-and-wait ambusher of small active prey. The adults barely

Left: *A new species in a new genus, the rare Three-horned pitviper (*Triceratolepidophis sieversorum*) was only discovered in Vietnam in 1999.*

Below: *Wagler's temple viper (*Tropidolaemus wagleri*) may be seen in large numbers in the Snake Temple, Penang. Although venomous, it does not seem to bite pilgrims or tourists.*

Opposite: *The Mangshan pitviper (*Zhaoermia mangshanensis*) may be on the verge of extinction. This strange viper is reputed to spit venom.*

moving for days. Juveniles are more active, even venturing into tall grass to hunt.

Wagler's pitvipers are noted for their docility and are reluctant to bite even when handled, in contrast to the more irritable bamboo pitvipers. Specimens in the Buddhist Snake Temple of Penang are allowed to roam freely, leading to their alternative name 'temple pitviper'. Dozens of snakes lie draped around the temple interior, reportedly soothed into compliance by the incense. They amaze tourists who are photographed holding them with apparent impunity.

Of all the arboreal Asian pitvipers, this is also one of the most popular in captivity, although it can be quite difficult to maintain since their inoffensiveness often makes them poor feeders. Despite their wide distribution and popularity in captivity, and the fact that snakebites must occur, the clinical effects of Wagler's pitviper bites remain poorly documented.

Range: Thailand, Malaysia, Philippines and Indonesia as far as Sulawesi • Max. length: 0.8–1.3m • Venom: Unknown despite potential for bites in the wild and captivity; no fatalities known • Habitat: Lowland forests and swamps but higher elevations where habitat is moist • Prey: Birds, frogs, lizards and rodents • Reproduction: Viviparous; 15-41 neonates • Similar species: Madura pitviper (*Tropiodolaemus huttoni*) known from two specimens collected in the 1940s from the Wavy Mountains in Tamil Nadu, southern India.

Mangshan pitviper *Zhaoermia mangshanensis*

A yellow-green terrestrial snake with dark brown irregular saddle-markings and also known as ironhead pitviper, this snake is named in honour of the eminent Chinese herpetologist, Er-mi Zhao. The only non-cobra reported to spit venom, it is extremely vulnerable, having a very small range and low population numbers. Recent human interference and destruction have pushed this unique pitviper to the edge of extinction.

Range: Mangshan, Nan Ling mountains, Hunan, China • Max. length: 1.4–1.7m • Venom: Nothing known; unconfirmed fatalities reported • Habitat: Montane forest from 700–1300m • Prey: Rodents, frogs and insects • Reproduction: Oviparous; 13–21 eggs.

Three-horned pitviper *Triceratolepidophis sieversorum*

A single specimen of a strange three-horned pitviper, representing not only a new species, but an entirely new genus, was discovered in the Annam mountains of Vietnam in 1999. It had been collected by a local medicine man and preserved in rice wine. The Vietnamese three-horned pitviper was so named because of the unusual arrangement of three raised horn-like projections and keeled scales. The first specimen, a male of 1.25m, was collected in an area of high snake diversity on the Vietnamese/Laotian border region. A second, female, specimen, was captured alive in the same region in 2001. The tongue-twister generic name means 'three-horn-scaled-snake'.

Range: Annam Mountains, Vietnam-Laos border • Max. length: 1.0–1.2m • Venom: Believed to be haemorrhagin and cytotoxin, but otherwise unknown; no snakebites known • Habitat: Karst forest • Prey: Nothing known • Reproduction: Nothing known.

AUSTRALASIA

Australasia comprises Australia, New Zealand, New Caledonia, New Guinea, and their surrounding archipelagoes. New Zealand has no terrestrial snakes and New Caledonia no venomous terrestrial snakes.

The unusual mammals and birds of this region, from kangaroos to kiwis, are well known, but the reptiles are pretty special too. There are at least 32 species of monitor lizards (Varanidae) in the region compared with approximately a dozen in Asia and Africa combined, and there were once many more species including *Megalania*, an extinct 5.7m monitor lizard which would have dwarfed today's champion lizard, the 2.7m Komodo dragon (*Varanus komodoensis*). The nonvenomous pythons (Pythonidae) are also represented by over 20 species compared with less than a dozen in Asia and Africa, but it is in the venomous Elapidae family that reptile speciation has really taken Australasia by storm. Australia and New Guinea contain over 100 species of terrestrial venomous snakes between them. Another three species are found on Bougainville and Buka Islands, and in the Solomon Islands, to the east of New Guinea, with a further isolated elapid on Fiji. Although Fiji falls outside the strict boundaries of Australasia it has been included here for convenience.

Elapids have been so successful in Australia and New Guinea because although many habitats and niches exist for snakes – there are jungles, swamps and deserts to be inhabited by small, leaf-litter dwelling, insect or frog-eating snakes, short-squat, cryptic-patterned, sit-and-wait ambushers, and sand-swimming, smooth-scaled lizard-eaters – there is no other competition. Vipers and colubrid snakes are extremely well represented elsewhere around the world, yet there are no vipers and very few colubrid snakes within Australasia and New Guinea, leaving the field clear for members of the Elapidae.

Right: *The Coastal taipan (*Oxyuranus scutellatus scutallatus*) hunts rats and bandicoots in the Queensland sugar-cane fields. It is highly venomous and fast moving. Snakebites constitute serious medical emergencies.*

Australian Elapids

Australia is truly the land of the elapids since they are the dominant snake family with over 90 species found in all habitats across the continent. Some are extremely dangerous but many are virtually harmless to man despite being technically venomous. Elapids are found everywhere in Australia, the only three snakes occurring on Tasmania being members of this family. Despite the high number of dangerous species, snakes only cause 1–2 deaths each year in Australia.

Whipsnakes

The eight Australo-Papuan whipsnakes are a classic example of venomous snakes that resemble nonvenomous species. In Europe or North America the diurnal (day active), fast moving, large-eyed predators of lizards and rodents are harmless colubrid snakes. Similar snakes are found in Australia and southern New Guinea – highly alert and active, big-eyed and fast-moving – but they are elapids with front-fangs and venom glands.

Lesser black whipsnake *Demansia vestigiata*

The black whipsnakes are alert and very fast, frequently seen but difficult to capture unless cornered. In southern Papua New Guinea (PNG) villagers often mistake them for the rare, but highly venomous, Papuan blacksnake (*Pseudechis papuensis*). I have been called to several 'Pap blaks' that were nothing more than lesser black whipsnakes.

Range: Northern and eastern Australia and southern New Guinea • Max. length: 1.2–1.7m • Venom: Weak coagulant; few snakebites with localized effects • Habitat: Savanna and savanna woodland • Prey: Lizards (skinks and agamas) • Reproduction: Oviparous; 4–13 eggs • Similar species: Greater black whipsnake (*D. papuensis*) from northern Australia but not Papua, despite its scientific name.

"Driving across eucalypt savanna east of Port Moresby I met a 1.0m whipsnake crossing the road. It immediately stopped, turned and reared up to face my vehicle, before advancing towards me in a threatening manner. This turned out to be a foolish mistake on the snake's behalf, as it gave me time to get out and capture it. On another occasion the tables were turned. As I prepared to 'milk' a large whipsnake it turned and bit me on the thumb, chewing for a moment before I could remove it. Expecting the effects of the venom to be negligible, I was very surprised to experience intense burning pain for about twenty minutes. **Central and Western Provinces, Papua New Guinea.**

Yellow-faced whipsnake *Demansia psammophis*

The three subspecies of yellow-faced whipsnake constitute the most wide-spread Australian whipsnake, occurring in every mainland state, although it may warrant threatened status in Victoria. Its patterning is light, in contrast to that of the more northern black whipsnakes, and consists of reticulated olive or reddish body, unicolour red tail and a dark circle around the eye which extends as a stripe to the angle of the jaw.

Range: Eastern, southern and western Australia • Max. length: 1.0–1.1m • Venom: Weak coagulant; few snakebites with localized effects • Habitat: Forest, woodland and savanna • Prey: Lizards (skinks and agamas) • Reproduction: Oviparous; 3–9 eggs • Similar species: Olive whipsnake (*D. olivacea*) from northern Australia.

Brownsnakes

The seven species of brownsnakes are highly variably patterned, they are not always 'brown'. Diurnal and found in many habitats, they are also amongst the most dangerous snakes in Australia, responsible for most of the rare snakebite fatalities. Deaths occur a few minutes to an hour later, from cardiac arrest, renal failure or cerebral haemorrhage. Because local pain and swelling are absent, and fang marks indistinct, bites are often ignored. Headache is a very serious sign of a potentially fatal brownsnake bite. A defining characteristic of *Pseudonaja* is the absence of the unique Australo-Papuan elapid 'temporolabial scale', a large scale between the 5th and 6th supralabial (upper lip-scales).

Eastern brownsnake *Pseudonaja textilis*

Eastern brownsnakes are noted for their speed and aggression. An angry eastern brownsnake will open its mouth, hiss, flatten its neck into a narrow hood and raise its body in an 'S'-shaped curve, before rushing forwards to bite its aggressor. It is an extremely common and dangerous snake which causes extremely serious snakebites.

Range: Eastern Australia, isolated northwestern populations, possibly New Guinea • Max. length: 1.8–2.2m • Venom: Pre- and postsynaptic neurotoxins and procoagulant; serious, causes rapidly fatal snakebites • Habitat: Open wet country, especially near watercourses • Prey: Lizards, small mammals and birds • Reproduction: Oviparous; 8–35 eggs • Similar species: Coastal taipan (*Oxyuranus scutellatus*).

"Specimens tentatively identified as eastern brownsnakes, were collected from Papua New Guinea's Northern Province. in the years following WWII. In the absence of specimens from the southern Papuan savannas it was assumed these snakes were accidental introductions from Australia, as eggs transported in military equipment. However, serum samples from snakebite victims and a misidentified museum specimen suggest this species may actually be indigenous to New Guinea. Central and Northern Provinces, Papua New Guinea.

Opposite right: *Yellow-faced whipsnakes (*Demansia psammophis*) are fast moving, diurnal lizard hunters.*

Opposite left: *In New Guinea the Lesser black whipsnake (*Demansia vestigiata*) is often mistaken for the more venomous Papuan blacksnake.*

Above: *The Eastern brownsnake (*Pseudonaja textilis*) raises up in an S-shaped curve before launching an attack.*

Right: *The Western brownsnake (*Pseudonaja nuchalis*) is a widespread and variably patterned snake.*

Gwardar or Western brownsnake *Pseudonaja nuchalis*

The western brownsnakes or gwardar, is a complicated species. Some of the geographical variations may even constitute new species, or at least subspecies. It is equally nervous, but less aggressive, than the eastern brownsnake, but it is still a highly dangerous snake responsible for serious snakebites.

Range: Australia excluding east and southern coasts • Max. length: 1.4–1.6m • Venom: Pre- and postsynaptic neurotoxins and procoagulant; serious, causes rapidly fatal snakebites • Habitat: Arid and semi-arid savanna and woodland • Prey: Lizards, small mammals and birds • Reproduction: Oviparous; 9–38 eggs • Similar species: Dugite (*P. affinis*) from southern Western Australia.

Taipans

There are two taipan species, although the inland taipan was formerly known as the small-scaled snake and belonged to another genus (*Parademansia*) until 1981. Today the Australian Aboriginal name *taipan* is synonymous with 'large, highly venomous Australian snake', but the species was hardly known until the late 1940s. In a country coming to terms with the danger posed by deadly tigersnakes, blacksnakes, brownsnakes, copperheads and death adders in the south and east, and just developing antivenoms to combat their bites, the discovery of an even larger and more venomous snake in the north was an unpleasant surprise. In 1950, a 20 year old snake catcher, Keith Budden, was bitten as he bagged one of the first Queensland taipans collected for venom research. Budden died in hospital before the taipan reached the Commonwealth Serum Laboratories in Victoria and the taipan's terrible reputation was born.

Coastal taipan *Oxyuranus scutellatus*

The Australian coastal taipan is a brown snake with a pale 'coffin-shaped' head but taipans from southern New Guinea, which are recognized as a separate subspecies, are more variable, ranging from brown to black, without the pale head but usually with an broad orange stripe down the centre of the back. In Australia taipans are rarely encountered but in southern PNG this is a common snake and the major cause of the up to 25 snakebite fatalities annually. Taipans are slender, fast-moving snakes that prey on dangerous prey, rats and their marsupial equivalent, the bandicoot. In pursuit of their prey they have substituted speed for strength and could be injured if the prey is not killed quickly. Hence they have long fangs, for an elapid, and highly toxic, fast-acting venom. For all their potential the taipan is a shy snake, preferring escape to attack, but if cornered it becomes one of the most dangerous snakes in the world, delivering a rapid series of strike-and-release bites high on the body. The taipan is Australasia's black mamba.

Range: Northern and eastern Australia and southern New Guinea • Max. length: 2.0–3.4m • Venom: Pre- and postsynaptic neurotoxins, myotoxins and procoagulants; bites fatal without treatment • Habitat: Savanna and grassy hillsides, sugar-cane fields • Prey: Rats and bandicoots • Reproduction: Oviparous; 13–18 eggs • Similar species: Eastern brownsnake (*Pseudonaja textilis*) and king brownsnake (*Pseudechis australis*).

"I was driving one evening when I saw a boy running along the road towing a piece of string with something attached to the end. In any other country it would have been a toy car but this was a little snake. I stopped and, telling the boy he should not treat an animal in this fashion, I scooped up the reptile and got

back into my car. Switching on the car interior light I looked down at the little snake on my lap. I discovered it was a taipan, fortunately deceased. Port Moresby, Papua New Guinea.

"The presence of a pre-synaptic neurotoxin means that paralysis is irreversible, even with antivenom. Unless respiration is maintained artificially the victim will die of respiratory paralysis. There were only two working ventilators in the general hospital where I was based so some patients had to be artificially ventilated by their 'wantoks' (relatives) using black 'ambubags', for 2-3 days until the paralysis wore off. Some patients did not survive. Port Moresby, Papua New Guinea.

Inland taipan *Oxyuranus microlepidotus*

The inland taipan is found in the fairly unpopulated heart of Australia. The only snakebites recorded have occurred to herpetologists, two catching and two keeping the species in captivity, and none have ended fatally due to the excellent medical facilities available. This species was formerly known as the small-scaled snake and has also been called the fierce snake, yet it is actually quite an inoffensive snake. Inland taipan habitat consisting of dried river beds where it hunts rodents in the maze of cracks that appear as the soil dries and shrinks. Coloration ranges from yellow with black flecking to dark brown. The inland taipan possesses the most toxic venom of any known snake. An average coastal taipan bite has the capacity to kill 95,000 mice; that of the inland taipan could kill 220,000 mice. Although it is difficult to determine human lethality from these figures one method used is to extrapolate from a 20gm mouse to a 70kg human. That would produce the following capabilities per average bite: coastal taipan 27 humans, inland taipan 62 humans. Even allowing for the 'bucket-chemistry', the lethality of the venoms of these snakes can be seen to be high.

Range: Central Australia • Max. length: 1.8–2.5m • Venom: Pre- and postsynaptic neurotoxins, myotoxins and procoagulants; four bites known, no fatalities • Habitat: Arid semi-desert plains • Prey: Rats • Reproduction: Oviparous; 9–12 eggs • Similar species: Eastern brownsnake (*Pseudonaja textilis*).

Opposite: *The Papuan taipan (*Oxyuranus scutallatus canni*) can be distingished by its orange vertebral stripe. This is a juvenile specimen.*

Above: *The Inland taipan (*Oxyuranus microlepidotus*), also called the Fierce snake. Although probably the world's most venomous land snake, it is not particularly aggressive.*

Blacksnakes

Only three of the six species of blacksnake are entirely black. The most widespread species, the king brownsnake, is actually brown, while Collett's snake (*P. colletti*) is an attractive combination of black and red and Butler's snake (*P. butleri*) is black with yellow spots. As a genus, they are extremely widely distributed throughout Australia and New Guinea, and are only absent from southern Western Australia, Tasmania and northern New Guinea. Five species are oviparous but the red-bellied blacksnake (*P. porphyriacus*), the southern-most species of the genus from New South Wales and Victoria, retains its eggs until full-term and then lays them with membranous shells that hatch immediately. Although not strictly viviparous this sort of reproductive strategy is common in species from extreme latitudes.

King brownsnake *Pseudechis australis*

The most widespread venomous snake in Australia also occurs in southern New Guinea and has the ability to adapt to virtually every habitat type. Confusingly called king brownsnake, this species is not a true brownsnake and does not belong to genus *Pseudonaja*, even though its coloration is predominantly brown, red-brown or yellowish. This large elapid is also known as the 'mulga snake', a name that originates from the mulga grassland habitat with which the snake is frequently associated. A large king brownsnake is a dangerous adversary which will defend itself vigorously by spreading its neck in a long narrow hood before forming a raised arch and rushing forwards to strike rapidly. Although snakebites may be common the venom of the king brownsnake is not as toxic as that of true brownsnakes or taipan so deaths are rare. Although found in southeastern West Papua (Indonesia) its distribution in Papua New Guinea is uncertain. Specimens have been found in the border region near Bensbach but precisely how far east the species occurs has yet to be determined.

Range: Most of mainland Australia, except south and east coasts, and southern New Guinea • Max. length: 2.4–2.7m • Venom: Anticoagulants and myotoxins; snakebites common, but only one fatality known • Habitat: Forest to savanna and desert • Prey: Lizards, small mammals and frogs • Reproduction: Oviparous; 4–19 eggs • Similar species: Eastern brownsnake (*Pseudonaja textilis*) and coastal taipan (*Oxyuranus scutellatus*).

"I visited the Pilbara region of Western Australia to investigate stories of a large brown snake that reportedly raises itself up and hoods when encountered. Although locally referred to as the Pilbara cobra, it is well known there are no cobras in Australia. Herpetologist Brian Bush and I searched out the native snakes of the region and came to the conclusion that the Pilbara cobra was probably a king brownsnake. When I eventually captured a specimen it rewarded us with a display of hooded defiance. **Pilbara, Western Australia.**

Papuan blacksnake
Pseudechis papuanus

The taipan is the most dangerous snake known to exist in New Guinea. It is the most venomous species present and the most frequent cause of serious and fatal snakebites in the south of New Guinea. Yet it is the 'Pap blak' that people fear throughout the southern provinces of Papua New Guinea. In the Mekeo language it is known as *auguma* which means 'to bite again', a reference to its multiple stab-and-release bites, a pattern also true of the taipan. In some regions the taipan and Papuan blacksnake are considered male and female of the same species despite a number of obvious physical differences.

Fear of the Papuan blacksnake probably goes beyond the actual snake and originates from its mythical status. The Pap blak is the magic-man's chosen snake, the one he uses as a vehicle for murder or assassination. To the Kiwai people a powerful magic-man is known as the *Ove-devenar* or 'Blacksnake Man'. He has the power to either send a snake to bite someone, or to turn into a snake himself to bite someone. Anybody bitten by a blacksnake is considered to have been targeted by a sorcerer and to reverse the spell the victim's relatives will seek out another magic-man in preference to going to hospital. This delay in the early management of a serious snakebite usually proves fatal. My snake-handling skills earned me similar status and the Mekeo name *Gaigai Tauna*, literally 'the man who has power over snakes'. The downside was that other magic-men saw me as a rival and I experienced a couple of incidents that I can only put down to professional rivalry.

Papuan blacksnakes are becoming increasingly rare, due to habitat destruction, active persecution and accidental poisoning by the introduced and highly toxic cane toad (*Bufo marinus*). It may already be extinct in Port Moresby and Central Province and is just hanging on in Western Province, Papua New Guinea, and neighbouring southern West Papua, Indonesia. I believe this dangerous, but unique and threatened, species should receive official protection.

Range: Southern New Guinea • Max. length: 2.0–2.4m • Venom: Neurotoxins, pro- and anticoagulants; snakebites are rare but capable of causing fatalities • Habitat: Savanna and savanna woodland, especially near water • Prey: Frogs, small mammals, lizards and birds • Reproduction: Oviparous; clutch size unknown • Similar species: Lesser black whipsnake (*Demansia vestigiata*).

Opposite: *Although brown in colour, the widespread King brownsnake (*Pseudechis australis*) is actually a 'blacksnake'.*

Above: *The Papuan blacksnake (*Pseudechis papuanus*) is becoming rare in Papua New Guinea due to habitat loss and possibly also ingestion of introduced and highly poisonous cane toads.*

"The Papuan blacksnake is rarely encountered in Papua New Guinea. I have only found four specimens: one was a severely damaged 2m specimen that I removed from a village school latrine and had to euthanize due to the severity of the machete wounds it had received, another was a recently killed specimen in the same village 16 years later, and two live specimens I obtained from Weam and Bensbach. These specimens were very different in appearance; I suspect there may be more than one species of 'Papuan blacksnake'. I have also examined a preserved Port Moresby specimen that has the scalation and body shape of *P. papuanus* but the coloration of *P. australis.* At the start of the twenty-first century I believe there is much still to learn about the dangerous snakes of New Guinea, just as there was about Australia's dangerous elapids at the start of the twentieth century. **Western Province, Papua New Guinea..**

Australian Copperheads

The name 'copperhead' is normally associated with woodland pitviper from eastern USA but the copperheads from southeastern Australia and Tasmania, are elapids, much larger, much stronger and potentially more dangerous.

Lowland copperhead *Austrelaps superbus*

Copperheads are copper-brown, powerfully built snakes with narrow pointed heads and an aura of strength and endurance. Being southern snakes from cooler climates, they are live-bearers which gives them a distinct advantage over egg-laying snakes in the same environment. Females can move around from sunny spot to sunny spot, basking to bring-on their offspring, and take shelter during cold conditions, whereas eggs, once laid, are immobile and subject to the vagaries of the weather and liable to perish in cold weather. The lowland copperhead is one of only three species of snakes in Tasmania. Range: Southeast Australia, Victoria, Tasmania, islands in the Bass Straits • Max. length: 1.0–1.7m • Venom: Postsynatic neurotoxins; a dozen snakebites, one fatality • Habitat: Low-lying grassland and scrub, often associated with watercourses • Prey: Frogs, lizards, small mammals, snakes • Reproduction: Viviparous; 9–45 neonates • Similar species: Highland copperhead (*A. ramsayi*) from Victoria and New South Wales; Pigmy copperhead (*A. labialis*) from Kangaroo Island, South Australia.

"I was searching for Tasmanian tigersnakes (*Notechis ater humphreysi*) in an area of dense coastal vegetation in northern Tasmania. In a few hours I captured four tigersnakes and two copperheads. My initial impressions of the two specimens were very different. The tigersnakes, although highly venomous, were easy snakes to pin down and secure, rather like Asian cobras, but the copperheads were different and won my instant admiration. When I tailed them, and attempted to pin the neck with a snake-hook, they reacted with strength and speed. The largest specimen, at about 1.5m, easily shrugged off the hook and surged for a nearby bush and I had to react quickly to recapture it. It took several attempts to secure this snake and even when I held it behind the head I was impressed by its tenacity and strength as it attempted to turn and bite me. It reminded me of an African snouted cobra. North Downs, Tasmania.

Slender Crowned Snakes

The four species of slender crowned snakes of genus *Drysdalia*, and the closely related short-nosed snake, *Elapognathus minor*, are small, innocuous species from the extreme southern coastal region of Australia from Perth to Tasmania and Sydney.

White-lipped snake *Drysdalia coronoides*

The white-lipped snake is brown or green with a bright white stripe, edged with darker pigment, running from the snout, under the eye to the neck. This is the most cold adapted of all Australian snakes, being found even in the centre of Tasmania and the cold highland regions of Victoria and New South Wales. The preferred habitat is tussock grass or sedgefield, where the little snake may be found sheltering under logs and in dense vegetation. Being a cold habitat species, it is curious that the white-lipped snake is an egg-layer but possibly its small size and secretive habits enable it to survive in conditions not usually conducive with oviparity. The conditions in Tasmania do affect the fecundity of this small snake, females only being able to reproduce every 2-3 years, as opposed to every year on the mainland. This little snake may be Australia's southernmost snake species since it even occurs on the small, barren islets off the southeastern coast of Tasmania.

Range: Southeast Australia, New South Wales, Victoria, Tasmania and islands in the Bass Straits • Max. length: 0.5m • Venom: Myotoxic or neurotoxic; one snakebite known, localized effects • Habitat: Grassland and woodland fringes • Prey: Skinks, their eggs and frogs. • Reproduction: Oviparous; 2–10 eggs • Similar species: Crowned snake (*D. coronata*) from southern Western Australia.

"It was so cold at Cradle Lake, central Tasmania, that I had to wear a coat, yet searching under pieces of discarded timber in a sheltered sedge and moss bed I quickly turned up four specimens of white-lipped snake. None showed any aggression and it was difficult to accept that they were front-fanged venomous snakes, members of the family Elapidae. Cradle Lake, Tasmania.

Opposite: *The Lowland copperhead (*Austrelaps superbus*) is a large and powerful, cold adapted species from southeastern Australia.*

Above: *The White-lipped snake (*Drysdalia coronoides*) is the southern-most snake in Australia, even occuring on small islets off southeastern Tasmania.*

Marsh Snakes

The two species of *Denisonia*, and two species of *Hemiaspis*, from Queensland and New South Wales are small, fairly innocuous live-bearing snakes associated with aquatic habitats. The rough-scaled snake is also a live-bearer associated with aquatic habitats but it is a much more dangerous species.

De Vis' banded snake *Denisonia devisi*

De Vis' banded snake is short and stout and patterned with light and dark brown bands. It lives and hunts for prey under fallen logs and in the deep mud cracks that form when watercourses dry out. Secretive by nature, most snakebites have occurred to herpetologists handling specimens. De Vis was a former curator of the Queensland Museum.

Range: Eastern Australia, north New South Wales and south Queensland • Max. length: 0.5m • Venom: Not known; snakebites recorded with pain and localized effects • Habitat: Open and riverine woodland • Prey: Frogs and geckos • Reproduction: Viviparous; 3–9 neonates • Similar species: Ornamental snake (*D. maculata*) from coastal Queensland.

Red-bellied swamp snake *Hemiaspis signata*

The red-bellied swamp snake may be black, grey or olive-green, resembling the dangerous rough-scaled snake, *Tropidechis carinatus*, which occurs in the same habitats. However, the swamp snake has smooth, rather than rough keeled scales. Secretive by nature, it shelters under logs and vegetation.

Range: Eastern Australia, coastal Queensland and New South Wales • Max. length: 0.8m • Venom: Not known; snakebites recorded with pain and localized effects • Habitat: Marshes and watercourses in forests and grasslands • Prey: Skinks and their eggs, and frogs • Reproduction: Viviparous; 3–20 neonates • Similar species: Grey snake (*H. damelii*) and rough-scaled snake (*Tropidechis carinatus*).

Above: *Red-bellied swamp snake (*Hemiaspis signata*).*

Left: *De Vis' banded snake (*Denisonia devisi*).*

Opposite top: *The Rough-scaled snake (*Tropidechis carinatus*) is highly venomous despite its inoffensive appearance.*

Rough-scaled snake *Tropidechis carinatus*

The rough-scaled snake is an accident waiting to happen. It is probably the Australian elapid most likely to be mistaken for a harmless snake, in this case the harmless common keelback (*Tropidonophis mairii*), which inhabits freshwater aquatic situations from New South Wales, around the coasts of Queensland and Northern Territory to northern Western Australia. Most elapids have smooth scales, or slightly keeled scales like the death adders, *Acanthophis* spp. It is highly unusual for elapids to have keeled scales like the rough-scaled snake, the only other example being the South African rinkhals spitting cobra, *Hemachatus haemachatus*.

The rough-scaled snake inhabits the same habitats as the harmless keelback, has keeled scales, is patterned a similar brown with darker crossbar pattern and also possesses a rounded, keelback-like head with large eyes and round pupils. Unfortunately, unlike the keelback it is highly venomous and rather irritable. Accidents result when persons familiar with keelbacks indiscriminately handle rough-scaled snakes. At 5mm, the fangs are relatively long for an elapid of 1m, and the fast-acting venom can cause unconsciousness within five minutes. Ignored snakebites may easily terminate fatally, though there is only one death on record, a party reveller who was bitten when he misidentified a rough-scaled snake for a juvenile carpet python (*Morelia spilota*). Carpet pythons have fragmented head scales, rather than the large scutes of the rough-scaled snake, which can be further distinguished from the common keelback by the absence of the loreal scale, present in keelbacks. Rough-scaled snakes are agile and may be found sheltering in arboreal ferns or tree holes. They are viviparous and usually nocturnal in habit, while keelbacks are oviparous and diurnal.

Range: Eastern Australia, north New South Wales and extreme south Queensland • Max. length: 0.8–1.0m • Venom: Pre- and postsynaptic neurotoxins and procoagulants; many snakebites with one fatality • Habitat: Coastal swamps and watercourses • Prey: Small mammals, frogs, birds and lizards • Reproduction: Viviparous; 5–18 neonates • Similar species: Common keelback (*Tropidonophis mairii*) from coastal east and north Australia.

Below: *Rough-keeled snakes may be mistaken for the harmless Common keelback (Tropidonophis* mairii*), below.*

Tigersnakes

Tigersnakes are among the best known of Australian venomous snakes and were probably the first encountered by European colonists. Prior to the advent of antivenom, tigersnakes were responsible for very large numbers of snakebite fatalities, but since the 1950s deaths have decreased to almost zero. They are temperate snakes from cool, damp habitats along the southern and southeastern coastal and highland regions of Australia, which were undoubtedly more widespread when sea levels were lower during glacial times. Raised sea levels isolated not only the Western Australia population from those of South Australia and Victoria-New South Wales, but also left populations high and dry on Tasmania and the islands of the Bass Straits and Spencer Gulf, South Australia. The island populations were given subspecific status, but whether they were subspecies of the eastern tigersnake or black tigersnake, or even whether subspecies were valid, has been disputed. The main characteristics separating the two species are geography and the midbody dorsal scale count, which is usually 17 in the black tigersnake and 19 in the eastern tigersnake, although individuals from both species may possess from 17 to 21 scale rows.

Eastern tigersnake *Notechis scutatus*

The eastern tigersnake is confined to the southeast Australian mainland population. The Perth population is now treated as a subspecies of the black tigersnake. It is from the eastern tiger that the name 'tigersnake' originates because it is usually banded dark and light brown with yellow flecking in the skin, particularly visible on the lower flanks. When disturbed the eastern tigersnake will flatten its neck and anterior body into a rudimentary horizontal hood, which is presented laterally to the threat. It strikes rapidly and vigorously and can inflict a serious snakebite, although it prefers flight to fight.

Range: Southeastern Australia, South Australia to New South Wales and Victoria • Max. length: 1.2–2.1m • Venom: Pre- and postsynaptic neurotoxins, myotoxins and procoagulants; snakebites common but few deaths due to antivenom availability, historically fatal outcome was common • Habitat: Rainforest to river flood plains • Prey: Frogs, lizards, birds, small mammals and fish • Reproduction: Viviparous; 14–80 neonates • Similar species: Western tigersnake (*N. ater occidentalis*) from Western Australia, which is also banded and was formerly a subspecies.

Below: *The Eastern tigersnake (*Notechis scutatus*): so named because many specimens are yellow and brown or black.*

Opposite: *Not all Black tigersnakes (*Notechis ater*) are black. This Tasmanian specimen is yellow.*

Black tigersnake *Notechis ater*

Five subspecies of black tigersnakes may be recognized from South Australia, southwestern Western Australia, Tasmania, and on islands in the Bass Straits between Tasmania and the mainland. Apart from the western tigersnake, most of the other subspecies are black in colouration. Being black, or melanistic, is a useful adaptation for life in cooler conditions because it enables the snake to warm quickly when basking to become active or digest a meal, and it will speed up the development of a female's embryos. It is common for banded specimens to occur within black populations and vice versa. The island tigersnake populations, with their specific lizard or sea-bird diets, beautifully demonstrate the island biographical characters of gigantism and dwarfish which may be observed in many other animal groups from tortoises to birds around the world.

Range: Southern Western Australia, Tasmania and islands of Bass Straits and South Australia • Max. length: 1.5–2.4m • Venom: Pre- and postsynaptic neurotoxins, myotoxins and procoagulants; snakebites common but few deaths due to antivenom availability, historically, a fatal outcome was common • Habitat: Marshland to grassland and coastal habitats, mainland and island locations • Prey: Frogs, small mammals, lizards, snakes and birds, often only one prey species available • Reproduction: Viviparous; 6–90 neonates.

"Tigersnake expert Terry Schwaner and I visited Chappell Island to film the giant 2.0m+ tigersnake that survives there on a diet of mutton-bird chicks. Mutton-birds return to Chappell Island, from the Arctic, each year to nest in underground burrows. Their large chicks form the staple diet of adult tigersnakes since there are no mammals present. The tigersnakes, which grow very slowly and can live to be over 30 years old, are faced with a feast or famine situation. Chappell Island tigers feed well for a few weeks and lay down sufficient reserves to last the ten months until next year's mutton-bird breeding season. The Chappell Island tiger population is very adult-heavy. The 45 tigersnakes we captured included only three juveniles. Young tigersnakes feed on skinks, of which there are six species on Chappell Island, but these lizards are not common and most juveniles probably do not survive to maturity. If the Chappell Island population is old and past reproductive age it may eventually become extinct. **Chappell Island, Bass Straits, Australia.**

Death Adders

Vipers are absent from Australasia, yet the biological niche for a short, squat, nocturnal, sit-and-wait ambusher still exists in a variety of habitats throughout most of Australia, New Guinea and the eastern archipelagos of Indonesia. Death adders may resemble vipers but they are truly members of the Elapidae that have evolved to occupy what is, elsewhere, a viper niche. Other viperine characteristics displayed by death adders include weakly or strongly keeled scales, raised horn-like scales over the eyes, which have vertically elliptical pupils, on an angular adder-like head. The fangs are fairly long and moveable, for an elapid, the patterning is cryptic and the flattened tail tip may be a contrastingly yellow or white, and used for 'caudal luring' prey within strike range. Death adders are the only elapids to adopt this hunting tactic. They also share live-bearing with most viper species, although this trait is not rare in Australasian elapids from the higher latitudes and altitudes, despite its rarity in non-Australasian elapids. The result is a good example of convergent evolution whereby the Australasian death adders have evolved along parallel lines to those followed by small terrestrial true vipers and pitvipers in the Americas, Africa, Europe or Asia.

The name death adder is thought to be a derivation from the early colonial name 'deaf adder', a reference to the fact that while other Australian snakes fled when approached this small squat snake stayed where it was and was presumed to be deaf. Of course all snakes are deaf to airborne sound, though they detect vibrations through the ground. The name does not relate to their killing capacities, even though many death adder bites terminated fatally prior to the development of an antivenom. Today, the effects of death adder post-synaptic neurotoxins can be rapidly reversed by antivenom or even with anticholinesterase drugs, for example, neostigmine. Death adders were once all included in a single catch-all species, now known as the southern death adder (*Acanthophis antarcticus*). Five species are now recognized, although as many as 11 taxa have been suggested by some authors. It is very likely that additional valid species will be described based on DNA analysis as much as their morphological differences.

New Guinea death adder *Acanthophis laevis*

In New Guinea death adders are widespread in virtually every habitat from lowland gardens to highland coffee plantations and pristine rainforests, from sea level to altitudes of 1800m in the heavily populated highlands. A dwarf race is reported from Eastern Highlands Province, while a giant 1m race is said to exist in the Markham River Valley. I can vouch for the fact that many Papua New Guineans cannot distinguish between death adders and ground boas (*Candoia aspera*), a harmless, stout, short-tailed, keeled-scaled snake which achieves 1m in length. For this reason I will remain sceptical about the validity of the giant Markham River death adder stories until I see a specimen.

Within New Guinea there is a considerable degree of variation in head shape, body and head scale roughness, coloration and patterning, and it is likely that there are several species masquerading as this single species. Although death adders probably pose no serious snakebite risk in Australia today, they may still be the cause of serious snakebites and fatalities in New Guinea, and it is probably the most dangerous snake species once one moves into the mountains or onto the north coastal lowlands of New Guinea and away from the range of the taipan.

Range: New Guinea and islands of eastern Indonesia • Max. length: 0.3–0.5m • Venom: Postsynaptic neurotoxin; many snakebites, few fatalities • Habitat: Woodland and cultivated gardens • Prey: Lizards and small mammals • Reproduction: Viviparous; litter size not known • Similar species: Northern death adder (*A. praelongus*) from northern Australia and southern New Guinea, and harmless New Guinea ground boa (*Candoia aspera*).

Desert death adder *Acanthophis pyrrhus*

The hot and arid Centralian deserts of Australia lie at the opposite end of the habitat spectrum to the moist, cool, coffee-cultivated slopes of highland New Guinea, which are inhabited by the New Guinea death adder. The desert death adder is the most specialized member of the genus, being more slender-bodied and heavily keeled that other species. It is sand-coloured with a black tail tip. Desert death adders are rarely seen during the day, preferring to shelter at the base of grassy hummocks and only become active once the heat of the day has diminished.

Range: Western and central Australia • Max. length: 0.75m • Venom: Postsynaptic neurotoxin; no snakebites on record • Habitat: Desert and semi-desert • Prey: Lizards, especially small dragons (agamids) • Reproduction: Viviparous; 10–13 neonates • Similar species: Pilbara death adder (*A. wellsi*) from Western Australia.

"We were driving on sandy and stony desert tracks at night and found two death adders inside 20 minutes. Initially I though I had found the recently described Pilbara death adder but by closely examinating the head scales and counting the body scales I determined that they were desert death adders. **Pilbara, Western Australia**

Opposite: *New Guinea death adders (*Acanthophis laevis*) are a serious snakebite threat in village gardens.*

Above: *The desert death adder (*Acanthophis pyrrhus*) is patterned to blend into the desert sand.*

Small Australian Elapids

Australia is inhabited by many small elapids that occupy niches associated with colubrids elsewhere in the world.

Bardick *Echiopsis curta*

At first glance the bardick looks a little like a death adder. It is fairly stout with a broad viper-like head, vertically elliptical pupils and may have black and white lip markings, but its scales are smooth, unlike the slightly or strongly keeled scales of some death adder populations, it lacks raised scales over the eyes and its body-shape transition, between body and tail, is less abrupt than in death adders. Three populations exist, without subspecific status, in southern Western Australia, on the Eyre Peninsula, South Australia, and inland in the South Australia, New South Wales and Victoria border region. A secretive snake, usually found hiding under debris on dry soil or sand substrates, the bardick will flatten its body defensively when disturbed, but its bite is not considered dangerous to humans.

Range: Southern Australia • Max. length: 0.7m • Venom: Anticoagulant; one snakebite known with localized effects • Habitat: Sandy grassland and coastal habitats • Prey: Frogs and lizards. • Reproduction: Viviparous; 3–14 neonates • Similar species: Death adders (*Acanthophis* spp.) and Lake Cronin snake (*E. atriceps*) from Western Australia.

Spotted snake *Suta punctata*

The overall colour of the spotted snake is red or orange brown, with a series of hard spots on the back of the head and neck. This species has been shuffled around half a dozen genera and it is distinctly different from the nine other member of genus *Suta*. A semi-fossorial (burrowing) species, it inhabits deep mud cracks, where it will encounter its prey, geckos, skinks and blindsnakes. The Pilbara is separated from the main northern population from the Kimberley region of Western Australia to Northern Territory and southeastern Queensland. It is an inoffensive snake.

Range: North and west Australia • Max. length: 0.6m • Venom: Nothing known; no snakebites reported • Habitat: Open woodland and grassy plains • Prey: Small lizards and blindsnakes • Reproduction: Viviparous; 2–5 neonates.

Eastern small-eyed snake *Rhinoplocephalus nigrescens*

The eastern small-eyed snake is a grey, smooth-scaled snake with tiny eyes found in a variety of habitats where it prefers to shelter under fallen timber and rocky slabs. This is the darkest of six species of Australian small-eyed snake, others, including those that extend up into southern New Guinea, are light grey or red-orange with a dark vertebral stripe. The only known fatality from an Australian small-eyed snake was caused by this species. It was unusual since death was greatly delayed and resulted from a cardiac arrest ten days earlier. The early signs were intense muscle pain, which progressed to myoglobinuria, loss of use of the lower limbs and renal failure. The myotoxic venom appears to have a destructive effect on heart muscle with myoglobin from damaged muscle leading to the kidney failure. This case, if genuine, emphasizes the effects of ignoring snakebites from small species that were previously considered relatively harmless.

Range: Eastern Australia, Queensland to Victoria • Max. length: 0.9–1.0m • Venom: Myotoxins; occasional snakebites with one fatality • Habitat: Rainforest, woodland and rocky outcrops • Prey: Small lizards and snakes, and their eggs • Reproduction: Viviparous; 1–8 neonates • Similar species: Northern small-eyed snake (*R. pallidiceps*) from Northern Territory.

"While searching for the endangered broad-headed snake (*Hoplocephalus bungaroides*) we found seven small-eyed snakes sheltering under the flat 'bushrock' slabs of Hawkesbury sandstone used by the broad-headed snake. **Morton National Park, New South Wales.**

Golden crowned snake *Cacophis squamulosus*

The four species of Australian crowned snakes are easily recognized by their head markings which consist of white-edged dark crowns. The golden crowned snake is the largest species, the others being amongst the smallest of elapids. Crowned snakes are secretive species that hide under any available shelter and are probably too small to represent a snakebite threat.

Range: Eastern Australia, coastal Queensland to New South Wales. • Max. length: 0.8–0.9m • Venom: Nothing known • Habitat: Wet forest and coastal habitats • Prey: Small lizards, snakes, their eggs and frogs • Reproduction: Oviparous; 2–15 eggs • Similar species: Three other crowned snakes (*Cacophis* spp.) on the same coastline.

Above: *The Eastern small-eyed snake (*Rhinoplocephalus nigrescens*) looks inoffensive but bites may cause kidney failure.*

Above Right: *The Bardick (*Echiopsis curta*) resembles a smooth-scaled death adder.*

Naped Snakes

The five species of naped snakes are small, nocturnal snakes with light cross-bands on the nape of the neck, though these are obscured in some species.

Orange-naped snake *Furina ornata*

The second largest member of the genus, the orange-naped snake, or moon snake, is the most widespread species of naped snake. It is orange-brown in colour with an orange nape band across its dark head crown. It usually inhabits secretive microhabitats such as deep mud cracks or underneath fallen timber or rocks.

Range: Australia, northern half from Western Australia to southern Queensland • Max. length: 0.7m • Venom: Probably neurotoxins and procoagulants; no snakebites recorded • Habitat: Woodland to grassland • Prey: Lizards, mostly skinks • Reproduction: Oviparous; 3–6 eggs • Similar species: Red-naped snake (*F. diadema*) from eastern Australia.

Grey-headed snake *Furina tristis*

The only naped snake to occur outside Australia, in southern New Guinea, and it appears to be the species that most prefers moist conditions, monsoon forest and cultivated gardens. It is also the largest species and therefore the one most likely to cause serious snakebites. The scales are smooth and the eyes are small. The underside is immaculate white.

Range: Northern Queensland, Australia and southern New Guinea • Max. length: 0.9–1.0m • Venom: Probably neurotoxins and procoagulants; one snakebite recorded with localized effects • Habitat: Monsoon forest, coconut plantations and gardens • Prey: Lizards, mostly skinks • Reproduction: Oviparous; 6 eggs • Similar species: Yellow-naped snake (*F. barnardi*) and Dunmall's snake (*F. dunmalli*) from southern Queensland and harmless colubrid slate-grey snake (*Stegonotus cucullatus*).

Left: *The Grey-headed snake (*Furina tristis*) is a common and fairly inoffensive snake from southern New Guinea and northern Australia.*

Below: *The Orange-naped snake (*Furina ornata*) is sometimes referred to as the Moon snake.*

Desert Burrowers

Many of the smaller Australasian elapids are semi-fossorial (burrowing), they inhabit leaf litter or shelter under fallen logs or rocks in grassland, woodland or rainforests, but for small, secretive snakes in desert and semi-desert there is frequently limited ground cover. This is probably why Australia has evolved a wide variety of fossorial elapids with small eyes and snouts adapted for burrowing. Although the extensive desert regions and other arid habitats are perfect for true burrowing species these specialized, often banded, snakes are not confined to arid sandy environments.

Southern shovel-nosed snake *Simoselaps semifasciata*

The southern shovel-nosed snake is one of the most widespread of over a dozen shovel-nosed and related burrowing and banded snakes occurring throughout Australia. These are the only Australasian snakes that possess enlarged, pointed rostral (snout) scales to aid their burrowing capabilities and with the orange and black banding pattern, they bear a striking resemblance to the harmless and unrelated shovel-nosed snakes (*Chionactis* spp.) of the southwestern USA and northwestern Mexican deserts. This is another case of an ecological niche being occupied by a member of the dominant Australian Elapidae where it would be inhabited by a harmless colubrid elsewhere in the world. The southern shovel-nosed snake is not strictly a desert species but it does occur in southern arid locations where it adopts a subterranean existence.

Range: Western Australia and South Australia • Max. length: 0.3–0.4m • Venom: Nothing known, no snakebites reported • Habitat: Coastal dunes and inland scrubland and grassland • Prey: Lizard and snake eggs • Reproduction: Oviparous; 2–5 eggs • Similar species: Eastern shovel-nosed snake (*S. australis*) from eastern Australia and harmless colubrid North American shovel-nosed snakes (*Chionactis* spp.).

Top: *The Southern shovel-nosed snake (*Simoselaps semifasciata*) is only one of a dozen specialised desert burrowing snakes in Australia.*

Above: *The Eastern bandy-bandy (*Vermicella annulata*) is a burrower threatened by intensive farming techniques.*

Eastern bandy-bandy *Vermicella annulata*

Five species of bandy-bandys are now recognized, often from isolated locations such as the Pilbara, but they are all poorly known, with the most widespread and long established species, the eastern bandy-bandy, still being poorly documented in the wild. It is now believed that intensive farming methods and ploughing may be threatening the species' existence. In the most southern part of its range, northern Victoria, the bandy-bandy has not been reported for over 100 years. Fossorial (burrowing) in habit, bandy-bandys only venture above ground at night, after rain, to search for blindsnakes which have also been driven onto the surface. They are unmistakable snakes that cannot be confused with any other Australian snakes since they are banded black and white and have tiny eyes. The defensive display of the bandy-bandy is reported to consist of a series of raised body loops which display the underbelly markings.

Range: Eastern Australia from the Gulf of Carpentaria to northern Victoria • Max. length: 0.8m • Venom: Nothing known; no snakebites reported • Habitat: Rainforest to heathland • Prey: Blindsnakes and elongate skinks • Reproduction: Oviparous; 2–13 eggs • Similar species: Northern bandy-bandy (*V. multifasciata*) from Northern Territory.

Broadheaded Snakes

The genus *Hoplocephalus* contains three arboreal and saxicolous (rock-dwelling) species from coastal New South Wales and Queensland. Slender-bodied but muscular, they are perfectly adapted to their habitats. Although small, the three species are sufficiently venomous to be considered dangerous but they are also endangered, especially the Sydney broadheaded snake.

Broadheaded snake *Hoplocephalus bungaroides*

The slender, agile, black and yellowish broadheaded snake defends itself by raising the anterior part of its body into an 'S' curve and flattening its already broad head. Probably the most endangered venomous snake in Australia, it is certainly the one with the most limited distribution. It is found only on the Hawkesbury sandstone outcrops around Sydney but even in this habitat its requirements are tightly defined. It lives under flat slabs in rock-on-rock situations, never rock-on-soil. It is therefore confined to bare rock ledges and plateaus, often near the edge and away from soil-forming vegetation. The same habitat is home to the its prey, velvet geckos, *Oedura lesueurii*. Unfortunately flat slabs of Hawkesbury sandstone, 'bushrocks', are popular with gardening Sydney-siders. The collection of bushrock is illegal as is any disturbance to the snakes or their habitat, yet rock collection continues.

Range: Sydney, New South Wales • Max. length: 0.8–1.0m • Venom: Procoagulants, little known; few snakebites, no fatalities • Habitat: Hawkesbury sandstone outcrops • Prey: Lizards, especially velvet geckos, frogs and small mammals • Reproduction: Viviparous; 4–20 neonates • Similar species: Stephen's banded snake (*H. stephensii*) from coastal New South Wales and pale-headed snake (*H. bitorquatus*) from New South Wales and Queensland.

"Broadhead expert Jonathan Webb and I visited the Hawkesbury sandstone outcrops to the south of Sydney. The habitat had been disturbed, either by bushrock or snake collectors, and although we found nine broadheads the population still appeared to be under threat despite legal protection of the snakes and regulations outlawing bushrock collection. Visiting otherwise inaccessible sandstone outcrops by helicopter failed to locate any snakes or their prey, possibly due to bush-fires and the reptile's inability to re-colonise. The situation seems even more serious than we thought. **Morton National Park, New South Wales.**

Above: *Australia's most endangered snake is the Broadheaded snake (*Hoplocephalus bungaroides*) from the Sydney area.*

Opposite: *The New Guinea small-eyed snake (*Micropechis ikaheka*) is common in coconut husk plantations on Karkar Island, Papua New Guinea, where it often kills the workers.*

Snakes of Melanesia

The Melanesian islands to the north and northeast of Australia, New Guinea and the Solomon Islands, as well as Fiji, are also inhabited by venomous snakes belonging to the family Elapidae.

New Guinea Endemics

At least eight northern Australian elapids also occur in southern New Guinea. The island of New Guinea and its satellite archipelagoes are also inhabited by at least 15 endemic elapid species. These include the Papuan blacksnake (*Pseudechis papuanus*) and New Guinea death adder (*Acanthophis laevis*) which are New Guinea representatives of Australian genera. The remaining Melanesian elapids belong to endemic genera, confined to New Guinea and the Solomons.

New Guinea small-eyed snake *Micropechis ikaheka*

If there was ever a snake species I considered 'my species', then this dangerous, smooth-scaled, New Guinea endemic is it. Unrelated to the smaller Australian small-eyed snakes (*Rhinoplocephalus* spp.), it is found in swamps and along monsoon forest creeks. The local dialect name *ikaheka* means 'land eel', a reference to its semi-aquatic habitat preferences. A pale snake, varying from light brown to cream, with cross-bands of brown or red contrasting with its grey head and tiny dark eyes, it is common on volcanic Kar Kar Island, 20kms off the coast of Madang Province, PNG, where it inhabits discarded coconut husk piles. Small-eyed snakes appear less common on the mainland and are even rare in southern PNG but they have been recorded up to 1,500m altitude in the highland provinces. It will eat other snakes including ground boas (*Candoia aspera*) and its own species. It is greatly feared by the villagers and plantation workers

Range: New Guinea, West Papua (Indonesian New Guinea) and Papua New Guinea • Max. length: 1.5–2.0m • Venom: Postsynaptic neurotoxins, anticoagulants and myotoxins; snakebites with three fatalities • Habitat: Monsoon forest, rainforest, plantations and riverine or swamp habitats • Prey: Lizards, other snakes, small mammals, possibly frogs • Reproduction: Oviparous, clutch size unknown • Similar species: Solomons small-eyed snake (*Loveridgelaps elapoides*) from Bougainville and the Solomon Islands and nonvenomous ringed python (*Bothrochilus boa*) in New Ireland, though the small-eyed snake does not occur in any of these locations.

"We suspected this snake was capable of killing people but in a country with many dangerous species we had no proof. We had third-hand accounts that the victims had been bitten by the 'white snake' but lacked a dead snake or a reliable identification. Modern biomedical analytical techniques like ELISA, comparing venom I had collected in the field with serum samples from fatal snakebite victims, provided the proof and confirmed two fatalities. When 'milking' 1.5-2.0m small-eyed snakes I noticed they have a tenacious chewing bite. It seems likely that when this species bites a human it does so in much the same manner enabling large quantities of venom to enter deep into the wound. I have caught over 40 specimens to date. **Kar Kar Island, Madang Province, Papua New Guinea.**

Highland and Island Rarities

Southern New Guinea shows strong relationships with Queensland, Australia, but the highland provinces, northern coast and the eastern archipelagos are inhabited by secretive, little-studied New Guinea endemics. Apart from the small-eyed snake the elapids comprise three species of New Guinea crowned snakes (*Aspidomorphus*) and at least nine species of forest snakes (*Toxicocalamus*). The forest snakes are especially poorly known and very rare, two species being known from single specimens, and four species from two, three, five and eight specimens respectively.

Müller's snake *Aspidomorphus muelleri*

Müller's snake is a widespread New Guinea species, occurring in New Guinea to altitudes of 1,500m, and also in the Bismarck Archipelago but it is a rare snake. Usually brown, with a red-brown underside and a white stripe passing below the eye, it is thought to be a nocturnal inhabitant of leaf-litter in damp situations. It causes minor snakebites resulting in localized pain and nausea.

Range: New Guinea and islands to west as far as Seram and east to New Ireland • Max. length: 0.4–0.6m • Venom: Poorly known; one snakebite with localized effects • Habitat: Lowland rainforests and monsoon forests • Prey: Not known but suspected to be frogs or lizards • Reproduction: Presumed oviparous • Similar species: Striped crowned snake (*A. lineaticollis*) from eastern Papua New Guinea.

Spotted forest snake *Toxicocalamus spilolepidotus*

The spotted forest snake is only known from the Kratke Mountains of Eastern Highlands Province, PNG. Two specimens alone exist, in the National Museum of PNG and the American Museum of Natural History. This species is the most attractively patterned of the known forest snakes, every cream scale is black-edged, presenting a reticulated pattern. The tail is short and blunted with a sharp terminal spine, an adaptation for burrowing found in many fossorial snakes. The spine, when dug into the ground, enables the snake to force itself forwards through the substrate.

Range: Papua New Guinea, Kratke Mountains only • Max. length: 0.78m • Venom: Nothing known; no snakebites reported • Habitat: Montane rainforest • Prey: Believed to comprise soft-bodied invertebrates, e.g. earthworms, possibly frogs or reptile eggs • Reproduction: Oviparous; clutch size unknown.

Below: *The Spotted forest snake (*Toxicocalamus spilolepidotus*) is known from only two preserved specimens.*

Opposite top: *The Solomons small-eyed snake (*Loveridgelaps elapoides*) is locally called the 'Shark of the Jungle'.*

Opposite bottom: *The Solomons coralsnake (*Solomonelaps par*) has yet to be recorded from Bougainville Island.*

Solomon Islands Endemics

The Republic of the Solomon Islands is an archipelago to the east of New Guinea that comprises six large islands and numerous smaller islands and atolls. Bougainville and Buka are politically part of Papua New Guinea but biogeographically comprise part the Solomon Islands archipelago. These islands are home to three endemic elapids.

Solomons coralsnake *Salomonelaps par*

Although called a coralsnake, this species, also known as Guppy's snake, has relatively large eyes, is apparently diurnal, and grey, brown or reddish with only faint bands present. Common throughout the Solomon Islands, and also found on the island of Buka to the north of Bougainville, it is strange that this species has yet to be collected from Bougainville itself. Fairly inoffensive, it will hiss if disturbed but rarely bites.

Range: Solomon Islands and Buka Island, Papua New Guinea • Max. length: 0.7–1.2m • Venom: Nothing known; one snakebite reported without any effects • Habitat: Rainforest • Prey: Frogs, lizards, mostly skinks and small snakes • Reproduction: Probably oviparous; clutch size unknown.

Solomons small-eyed snake *Loveridgelaps elapoides*

This brightly patterned snake is white with regular broad orange and black bands. Named in honour of the herpetologist Arthur Loveridge (1891-1980), one local name is 'shark of the jungle' though quite why is a mystery. The preferred habitat of this rare nocturnal snake is rainforest leaf-litter, especially along streams. An expert on Solomons herpetology reports finding a small-eyed snake trying to swallow an similar sized coralsnake on Guadalcanal Island.

Range: Solomon Islands • Max. length: 0.8–1.0m • Venom: Nothing known; no snakebites reported • Habitat: Rainforest • Prey: Lizards, snakes, possibly frogs • Reproduction: Probably oviparous; clutch size unknown • Similar species: New Guinea small-eyed snake (*Micropechis ikaheka*) and the sea kraits (*Laticauda* spp.).

Bougainville & Fiji Endemics

The islands of Bougainville, between New Guinea and the Solomon Islands, and the island of Viti Levu, Fiji, are each home to one endemic elapid snake.

Bougainville coralsnake *Parapistocalamus hedigeri*

On Bougainville, the largest island in the Papua New Guinea far-eastern North Solomons Province, occurs one of the world's least known elapids. The Bougainville or Hediger's coralsnake, a small brown snake with a pale nape band, is known from very few specimens and almost no field data. It has not been found on any neighbouring islands. Semi-fossorial, it seems to be a nocturnal inhabitant of rotten logs and leaf mould and it would be thought an insignificant little snake but for one small fact. The presence of a diastema, a toothless gap, behind the fangs suggests that the closest relatives of the Bougainville coralsnake might not be the other Australasian elapids of New Guinea, Australia or even the Solomons. This little snake may represent an unknown subfamily.

Range: Bougainville Island, Papua New Guinea • Max. length: 0.3–0.5m • Venom: Nothing known; no snakebites reported • Habitat: Rainforest • Prey: Nothing known • Reproduction: Probably oviparous; clutch size unknown.

Fiji burrowing snake *Ogmodon vitianus*

The local names for this strange little snake include *bola* and *gata ni balabala* which translates to 'snake of the mountain glens'. A dark snake with a white nape band, it would be pretty unremarkable if it were not the only terrestrial venomous snake for thousands of kilometers. Fiji is located 3,000km east of Australia and a similar distance from the Solomon Islands to the northwest so the presence of this tiny front-fanged venomous snake is very interesting. Unfortunately, habitat destruction and introduced animals on Viti Levu, the only island where the snake has been found, are serious threats to its survival. The Fiji snake is a burrower that may occur as deep as 1.0m below the surface and is often only found on the surface after rain. The IUCN list this species as vulnerable but it probably needs stronger protection and active conservation to prevent its extinction.

Range: Viti Levu, Fiji • Max. length: 0.2–0.3m • Venom: Nothing known; no snakebites reported • Habitat: Rainforest and yam gardens in mountain valleys • Prey: Soft-bodied invertebrates, such as earthworms and soil arthropods • Reproduction: Oviparous; 2–3 eggs.

Above: *The Fiji burrowing snake (*Ogmodon vitianus*) is isolated on Fiji, 3,000kms from its nearest relative.*

Left: *The Bougainville coralsnake (*Parapistocalamus hedigeri*) is believed to be most closely related to American or Asian coralsnakes.*

Australasian Rear-fanged Colubrids

The large family Colubridae is poorly represented in this region and most rear-fanged colubrids in New Guinea and Australia are small and do not poses a snakebite risk. Even the brown treesnake, at home in New Guinea and the Solomons, is not a problem but when this voracious predator reaches a snakeless island like Guam it is a different matter.

Brown treesnake *Boiga irregularis*

The brown treesnake is an aggressive, nocturnal, arboreal predator of vertebrates. Widely distributed through New Guinea and the Solomons, is at equilibrium with its environment and other species. However, after WWII the brown treesnake reached Guam in the Marianas Islands, possibly as eggs inside military equipment returning from New Guinea/Solomons, to the USA. The brown treesnake found an island with no predators, only easy prey. They multiplied and preyed on Guam's flightless birds, fruit bat and indigeous lizards, pushing several species to extinction and others to the brink. They also caused numerous power-cuts by entering electric boxes and bridging terminals. Snakes grew to the size of small pythons on a diet of domestic chickens and eggs and began to cause worrying snakebites to children and babies, though so far without a fatality. Eradication methods attempted include traps and poison, even viruses have been considered. With an estimated one million snakes on Guam, containment may be a more logical course of action, using snake fences and patrols with lights and snake-sniffing dogs to prevent them entering cargo areas and being transported to neighbouring Saipan, the Hawaiian Islands or United States mainland.

Range: New Guinea and Solomon Islands, introduced to Guam. • Max. length: 1.0–2.3m • Venom: Venom composition not known; numerous bites with localized effects, no fatalities • Habitat: All habitats. • Prey: All small terrestrial vertebrates • Reproduction: Oviparous; 3–11 eggs • Similar species: Banded treesnake (*B. fusca*) from northern Australia.

"The largest treesnake I caught on Guam was almost 2.0m. It had eaten a chicken egg and was resting on the fence, metres from the cargo bays and it's ride to pastures new. **Guam, Mariana Islands.**

Below: *The voracious Brown treesnake (*Boiga irregularis*) has caused an ecological catastrope on the island of Guam.*

THE OCEANS

Seasnakes are fully marine, live-bearing snakes while sea kraits are amphibious and come ashore to digest meals and lay eggs. Fifty-eight seasnake species and six sea kraits are known, with four new species being described. Seasnakes occur in the Pacific and Indian Oceans but sea kraits are confined to the western Pacific. One seasnake is the most widespread snake in the world, from S. Africa to the America's Pacific coast.

Some marine snakes are generalists, found in many habitats feeding on a variety of prey, while others have specialized habitat and prey preferences. Three species feed on fish eggs and may be becoming nonvenomous. Most species inhabit inshore waters, estuaries and mudflats, or coral reefs, but some prefer open or deep water. Ten seasnake genera contain single species while three genera contain only two species. Thirty-six seasnakes are banded seasnakes (*Hydrophis*) with the pipe seasnakes (*Aipysurus)* and sea kraits (*Laticauda*) the next largest genera with six species each.

Seasnakes and sea kraits represent two elapid lineages that evolved to survive in seawater, their physiology and morphology adapting accordingly. The most obvious adaptation is the paddle-shaped tail for swimming, but several species compress their entire bodies like ribbons to enhance this oar-like effect. Since seasnakes do not leave the ocean they have no requirement for the large, overlapping belly scales which make land locomotion possible for sea kraits and both terrestrial snakes. Therefore these scales have become reduced in size so they are indistinguishable from the normal body scales. This allows greater lateral compression of the body for swimming but makes locomotion on land impossible.

The physiological adaptations achieved by seasnakes to enable deep diving on a single breath, surfacing without suffering from 'the bends', prevention of excessive salt intake, buoyancy control and obtaining drinking water, are incredibly complex (*see* page 23). Seasnakes are among the most specialized, yet least studied, of all snakes.

Right: *The orange banded New Caledonian sea krait (*Laticauda *sp. nov.) was formerly considered part of the widespread species Yellow-lipped sea krait (*L. colubrina*) but is on the verge of being recognised as a separate species.*

Sea Kraits

Sea kraits are often mistaken for true seasnakes because of their paddle-shaped tails, but they represent a separate lineage of marine snakes. Unlike seasnakes they come onto land to mate, lay eggs, digest meals and drink fresh water. Six species are recognized in genus *Laticauda* with two more being described. Sea kraits are naturally preyed on by sea eagles, ospreys, reef herons and tiger sharks, but populations on islands where mongooses have been introduced have been pushed to extinction, they are easy prey and will not defend themselves.

Yellow-lipped sea krait *Laticauda colubrina*

This widespread sea krait is an eel specialist that refuses all other prey and may coexist alongside a fish specialist like Crocker's sea krait. In Fiji the sexes even feed on different species of eels, the males preying on smaller species than the females. Although venomous, sea kraits are completely inoffensive to humans and it is possible to pick up handfuls of snakes without any attempting to bite as I found when catching sea kraits in New Caledonia. This is the most agile and terrestrial of the sea kraits, often traveling to the centre of small islands, scaling cliffs or climbing low bushes. On Rennell Island, in the Solomon Islands, the yellow-lipped sea krait crawls over land to enter the land-locked brackish Lake Te'Nggano, already inhabited by three native species; the endemic Crocker's sea krait (*L. crockeri*), an eel and a goby. The native sea krait feeds on the goby while the invasive yellow-lipped sea krait preys on the eel.

Range: Eastern India to western Pacific • Max. length: 1.10–1.4m • Venom: Postsynaptic neurotoxins and myotoxins resulting in renal failure; snakebites rare but known • Habitat: Coral reef and inland • Prey: Eel specialist, especially moray, zebra and conger eels • Reproduction: Oviparous; 4–20 eggs.

Brown-lipped sea krait *Laticauda laticaudata*

The brown-lipped sea krait differs from the yellow-lipped sea krait in more than lip scale colour. It is also deeper blue

between the black rings, and is more slender with a longer tail. The ranges of the two species overlap though most of their ranges but it usually confines its terrestrial activity to the beach.

Range: Eastern India to western Pacific • Max. length: 0.90–1.0m • Venom: Postsynaptic neurotoxins and myotoxins resulting in renal failure; no snakebites known • Habitat: Coral reef and beaches • Prey: Eels, especially moray eels, and elongate fish, also gobies in their burrows • Reproduction: Oviparous; 1–14 eggs.

“Zebra eels strongly resemble sea kraits and may obtain some protection from mimicking a venomous snake, but when we placed a zebra eel in a large tank with a newly captured sea krait the snake found and ate the eel during the first night. Madang, Papua New Guinea.

“I heard a snake moving over the dead leaves and when I looked towards the sound and saw the snake I instantly thought ‘what is a Californian kingsnake doing here?’ Seeing a banded snake in habitat more suited to terrestrial snakes made my mind do a geographical double-take. Sea kraits may be very common, in 48 hours on two islands I found 75 specimens of two species. I have also encountered sea kraits in southern Papua New Guinea and Indonesia. Ile Moro, New Caledonia.

Similar species: Erabu sea krait (*L. semifasciata*) from Indonesia to Japan, Niue sea krait (*L. schistorhynchus*) from Cook Islands, Samoa and Tonga, Vanuatu sea krait (*L. frontalis*), and lake-dwelling Crocker's sea krait (*L. crockeri*) from Rennell Island in the Solomon Islands.

Opposite: *This Yellow-lipped sea krait (*Laticauda colubrina*), from Krakatau, Indonesia, can travel over land, unlike true seasnakes.*

Below: *The Brown-lipped sea krait (*Laticauda laticaudata*) lays eggs on land, unlike the live-bearing true seasnakes.*

True Seasnakes

The term 'true seasnakes' is used for all marine snakes with paddle-shaped tails excluding the sea kraits (*Laticauda*). There are 58 species known to science with at least two new species being described. Many are highly venomous but a few are evolving towards becoming nonvenomous. Some seasnakes are especially common, others extremely rare and poorly documented. Over half of the known species of true snakes belong to the genus *Hydrophis*, commonly known as the banded seasnakes, even though most seasnakes are either ringed or banded.

Olive seasnake *Aipysurus laevis*

The olive seasnake is the largest of a group of seanakes, known collectively as pipe seasnakes, which still possess broad ventral scales like land snakes. The home range of the olive seasnake may be quite small, sometimes consisting of a single coral bommie (outcrop). Leisure divers frequently encounter the same olive seasnake, in same place, day after day, dive after dive, and being an inquisitive species, it will investigate visitors closely. Attracted by the reflection of the glass it will swim directly into the face-mask, often to the dismay of the wearer. Although generally a docile species there have been instances when a rebuffed olive seasnake has pursued and attacked a diver, a dangerous situation since this is one of the few species which has 4.7mm fangs long enough to penetrate a neoprene wetsuit. Only tiger sharks will prey on large seasnakes.

Range: Tropical Australia, New Guinea and New Caledonia • Max. length: 2.0m • Venom: Postsynaptic neurotoxins and myotoxins; serious, non-fatal snakebites recorded • Habitat: Coral reef • Prey: Reef fish generalist, occasionally prawns • Reproduction: Viviparous; 1–5 neonates • Similar species: Seven smaller pipe seasnakes, mostly in Western Australian waters, but Eydoux's seasnake (*A. eydouxii*), which feeds entirely on fish eggs, extends north into Asian waters.

"In 2002, diving on Ashmore and Hibernia Reefs, we captured 120 seasnakes of seven species, including the two endemic pipe seasnakes, the dusky seasnake (*A. fuscus*) and leaf-scaled seasnake (*A. folisquama*). We also caught 30 olive seasnakes. Known as the 'Seasnake Capital of the World', these reefs are home to a greater diversity of seasnakes than even the Great Barrier Reef. Ashmore and Hibernia Reefs, Western Australia.

Below: *This large Olive seasnake (*Aipysurus laevis*) has flattened its body to increase its swimming ability.*

Annulated turtle-headed seasnake *Emydocephalus annulatus*
This snake has belly scales like a terrestrial snake, but it is not necessarily a primitive species since it has adapted and evolved in a different direction. The fangs are less than 0.15mm in length and the venom glands are also greatly reduced in size because they are not required to deal with its specialized diet of fish eggs.

In many seasnakes the female is larger than the male, or rarely, vice versa, but in the turtle-head the sexes are the same size. Yet the turtle-head does exhibit two differences between the sexes. Females are creamy-yellow with fragmented black saddle-markings, while males are black. In addition the rostral scale of the male bears a sharp, downward-facing, beak-like projection, which is absent in the round-snouted female. It was suggested that this projection is used to scrape fish eggs off the coral, but since both sexes feed on eggs its absence in the female seemed curious.

Range: Tropical Australia, New Caledonia and the Philippines • Max. length: 0.75m • Venom: Venom weak, presumed postsynaptic neurotoxic; harmless to humans • Habitat: Coral reef and shallow seagrass beds • Prey: Fish eggs of reef-dwelling blennies and gobies • Reproduction: Viviparous; 2–5 neonates • Similar species: Japanese turtle-headed seasnake (*E. ijimae*) from Ryukyu Islands and Taiwan.

"When we dived on Ashmore we found the annulated turtle-headed seasnake to be particularly common and sometimes encountered pairs swimming together. One of our cameramen filmed a dark male courting a cream female. As he swam alongside and above her, he stroked her neck with his snout projection. He was using it to court her in the same way that a male python or boa uses its pelvic spurs to stroke and court a female. Since females do not reciprocate, the snout projection is absent in females, just as the spurs may be reduced or absent in female pythons and boas. Ashmore Reef, Western Australia.

Above: *A pair of Turtle-headed seasnakes (*Emydocephalus annulatus*). The male is black and possesses a snout protection with which he courts the female by scratching the back of her neck.*

Stokes' seasnake *Astrotia stokesi*

Distinguished from all other seasnakes by its greatly reduced and paired ventral scales. This is not the longest seasnake, but it is probably the stoutest and heaviest. The broad head contains the longest fangs of any marine snake (up to 6.7 mm), so a bite can penetrate a 5mm neoprene wetsuit. A benthic (bottom-dwelling) species it preys on bottom-dwelling fish and rarely comes into contact with humans unless trawled from the depths by fishermen. Although it has a reputation for aggressive behaviour, often launching a repeated attack and injecting venom with every bite, Stokes' seasnake is not one of the eight species reported to have caused human fatalities. On one occasion thousands of Stokes' seasnakes have been seen drifting in 'slicks' many metres wide and several kilometres long in the Strait of Malacca, west of Malaysia.

Range: Pakistan to Taiwan and south to New South Wales, Australia • Max. length: 1.2–2.0m • Venom: Postsynaptic neurotoxins; serious, non-fatal snakebites recorded • Habitat: Deep water outer slopes of coral reefs • Prey: Goby and goby-type fish specialist • Reproduction: Viviparous; 1–12 neonates.

"We hoped to find Stokes' seasnake whilst filming seasnakes on Ashmore Reef, the 'Seasnake Capital of the World', with 14 resident species. Stokes's seasnake inhabits the deeper water outside the reef. Unfortunately for me this last dive went wrong and I ran out of air at a depth of approximately 22m, even though my gauge indicated that I still had sufficient air left for my ascent. I located the dive master and, sharing his air, we made a controlled ascent. I then spent two hours on oxygen to alleviate any risk of 'the bends'. Seasnakes do not seem to have these problems. Asmore Reef, off Broome, Western Australia.

Above: *Stokes' seasnake (*Astrotia stokesi*) is the largest and heaviest of seasnakes. Its fangs can penetrate a wetsuit.*

Opposite: *The Beaked seasnake (*Enhydrina schistosa*) causes snakebite fatalities to fishermen wading in estuarine water.*

Beaked seasnake *Enhydrina schistosa*

This species gets its common name from a soft beak-like rostral projection at the front of the upper jaw but a dagger-like scale under its chin will also identify it. Beaked seasnakes hunt their bottom-dwelling fish by touch in low-visibility estuarine waters. An extremely dangerous species responsible for the majority of serious and fatal seasnake bites, the venom yield of 8.5-79mg, is far greater than the estimated fatal human dose of 1.5mg. Although the beaked seasnake may not be the most venomous seasnake, it is a species that frequently comes into contact with humans in shallow estuaries where villagers fish and wade. It has been estimated that a full bite from a beaked seasnake, delivering the maximum dose of venom, has the capacity to kill 22 people. All seasnake antivenom is produced from the venom of this species. Female beaked seasnakes also produce the largest number of offspring of any seasnake suggesting that mortality of neonates is high.

Range: The Persian Gulf to Queensland, Australia • Max. length: 1.2m • Venom: Myotoxins and postsynaptic neurotoxins; most snakebite fatalities are due to renal failure • Habitat: Estuaries and river-mouth mud-flats • Prey: Catfish and pufferfish specialist but occasionally takes other fish, even squid • Reproduction: Viviparous; 4–34 neonates • Similar species: Zweifel's seasnake (*E. zweifeli*) from the Sepik River mouth, Papua New Guinea.

"Even in its ideal habitat this is not a common species. In four nights at Weipa we recorded 220 seasnakes, of eight species, including only ten beaked seasnakes. This is also not a very hardy species. A short time after capture the head and neck would swell grossly with oedema even if the snake was handled gently. It seems likely that because it inhabits a very 'soft world' of water and gently inclined mud flats, and never encounters hard objects, it is unable to withstand even slight head trauma. This has serious implications for maintaining beaked seasnakes in captivity for antivenom production, since they might not survive long in aquariums. **Weipa, Queensland, Australia.**

Short-tailed seasnake *Lapemis curtus*

The Short-tailed seasnake consists of an Asian subspecies, *L. curtus curtus*, and an Australasian subspecies, *L. c. hardwickei*, although some authors consider Hardwicke's seasnake a valid species. The Short-tailed seasnake is a large and powerfully built snake with a short tail and a broad head. Its body scales are hexagonal and those on the lower sides and under the chin may be keeled or tuberculate in males. An aggressive species with a large mouth, this species has caused human fatalities.

Range: Persian Gulf to Taiwan and south to Australia • Max. length: 1.0–1.5m • Venom: Postsynaptic neurotoxins and myotoxins causing renal failure; human fatalities recorded • Habitat: Clear coral reef waters to silty shorelines and turbid estuaries • Prey: Fish generalist that feeds on a wide variety of species, also crustaceans and squid • Reproduction: Viviparous; 1–15 neonates.

"When we spent four nights netting seasnakes in the estuary at Weipa, northwest Queensland, we found Hardwicke's seasnake to the most frequently encountered species, accounting for 78 per cent, 171 of 220 seasnakes observed or captured. Those specimens captured were aggressive, biting but failing to penetrate the neoprene diver's gloves were wore to handle them. Since we were collecting venom we had to handle them quickly to prevent them wasting venom by biting gloves, the boat and each other. Weipa, Queensland, Australia.

Below: *We found Hardwicke's seasnake (*Lapemis curtus hardwickei*), to be the commonest and most aggressive seasnake in the Hey-Emblem estuary at Weipa, northern Queensland.*

Horned seasnake *Acalyptophis peroni*

This snake gets its name from the raised horn-like spines over each eye. It also has angular lip-scales that present a zigzagged appearance from the front, and every scale of the body is keeled or rough to the touch. Demonstrating an ontogeneric switch in diet with increasing age, from shrimp to fish, horned seasnakes actively search burrows and coral crevices for prey. This seasnake is frequently coated with seaweed, which may indicate long periods of inactivity on the seabed. Despite possessing one of the most toxic venoms of any seasnake the venom yield is small and there have been no human bites recorded. Predatory sharks possess such a large body mass that a bite from a seasnake would have no effect, so they can eat horned seasnakes with impunity.

Below: *The only seasnake with horns,the Horned seasnake* (Acalyptophis peroni) *may become camouflagued by a covering of algae and seaweed.*

Range: Northern Australia to New Caledonia, New Guinea, Thailand, the Philippines and Taiwan • Max. length: 1.1–1.3m • Venom: Postsynaptic neurotoxins and myotoxins; no snakebites recorded • Habitat: Coral reefs, sandy, silty or seagrass beds • Prey: Specialist; juveniles prey on shrimps; adults feed on gobies and possibly blennies • Reproduction: Viviparous; 1–8 neonates.

"We found horned seasnakes on the surface at night, both in the silty estuary at Weipa and over the coral reef at Hibernia Reef but never encountered them during the day at either Hibernia Reef, or on the surface at Ashmore Reef. We did not dive Weipa because of treacherous currents, crocodiles and tiger sharks. This does not seem to be a commonly encountered species.
Weipa, Queensland and Hibernia Reef, Western Australia.

Yellow seasnake *Hydrophis spiralis*

Although rarely encountered, the yellow seasnake is the world's longest seasnake and since it is also one of the most aggressive species, inclined to bite at the slightest provocation, it must be considered highly dangerous. It has caused human fatalities. Little is known about its natural history other than it seems to prefer deep water, only being seen at the surface when it is basking to increase its body temperature.

Range: The Persian Gulf to the Philippines and Indonesia, one specimen from New Caledonia • Max. length: 1.4–2.74m • Venom: Postsynaptic neurotoxins and myotoxins; snakebites recorded, including one fatality • Habitat: Deep water down to 10m • Prey: Eels and other elongate fish • Reproduction: Viviparous; 5–14 neonates.

Elegant seasnake *Hydrophis elegans*

The relatively slender, elegant seasnake is one of the commonest seasnakes in Australian waters, especially in river mouths and bays where it hunts on the muddy bottom. It may also be Australia's longest seasnake. It is primarily found in northern waters but moves inshore and southwards in summer. Although snakebites have occurred, mostly to fishermen, there have been

Left: *The Elegant seasnake (*Hydrophis elegans*) is a common Australo-Papuan species in the largest genus of seasnakes, the banded seasnakes* Hydrophis.

no confirmed fatalities due to elegant seasnake bites.
Range: Australia and New Guinea • Max. length: 2.1m • Venom: Postsynaptic neurotoxins and myotoxins; non-fatal snakebites recorded • Habitat: Shallow, moderately turbid inshore waters over sand or silt • Prey: Primarily eels but also catfish and mullet • Reproduction: Viviparous; 8–23 neonates.

Small-headed seasnake *Hydrophis gracilis*

The posterior two-thirds of the small-headed seasnake's body are stout and powerfully built but the anterior third and the neck are extremely narrow, the head being so tiny and discrete that the species was once placed in a separate genus, *Microcephalophis*, meaning 'tiny-headed-snake'. The body proportions relate to this snake's diet of eels, which it captures by inserting its small head and neck deep down their burrows. It can dive to 30m in search of prey. No bites to humans are known for this species, possibly because its mouth is too small, but its venom is probably highly toxic.
Range: The Persian Gulf to India and Sri Lanka, also Hong Kong • Max. length: 1.0m • Venom: Postsynaptic neurotoxins and myotoxins; no snakebites recorded • Habitat: Bottom-dweller on sand or silt seabeds, often over 30m deep • Prey: Eel specialist • Reproduction: Viviparous; 1–16 neonates.

Lake Taal seasnake *Hydrophis semperi*

It would seem a misnomer to call this species a 'seasnake' since it is only found in the freshwater Lake Taal. It is poorly known and considered rare. A small species, it is also considered a local delicacy known as *duhol* so the population could be under threat.
Range: Lake Taal, Luzon Island, Philippines • Max. length: 0.5–0.75m • Venom: Postsynaptic neurotoxin and myotoxin; snakebites recorded, including a possible fatality • Habitat: Freshwater lake • Prey: Not known • Reproduction: Probably viviparous but otherwise undocumented • Similar species: Genus *Hydrophis* contains at least 36 species, mostly in marine habitats but some enter large rivers and swim upstream into brackish water and even into freshwater, such as the recently described Sibau River seasnake (*H. sibauensis*) from Kalimantan, Borneo.

"We found the elegant seasnake the commonest *Hydrophis* species in four nights of netting in the estuary, accounting for 20 of the 220 marine snakes encountered. Contrary to reports, they were not aggressive, in contrast to both beaked and Hardwicke's seasnakes, but some of the impressive adults were over 1.5m in length. **Weipa, Queensland, Australia.**

"In 1986 I obtained an unidentified *Hydrophis* from a spear fisherman some 60km up stream on the Oriomo River in freshwater conditions. Some of the banded seasnakes have been found swimming up tidal rivers so this was an interesting find. Sadly the specimen succumbed to its injuries and could not be preserved. **Old Zim, Western Province, Papua New Guinea.**

Arafura mangrove seasnake *Parahydrophis mertoni*

Most seasnakes are found in estuaries or over coral reefs but three species are associated with mangrove swamps and mud-flats where they forage for gobies and other small fish hiding in crab burrows. As the rising tide floods the burrows the snakes follow the tide to capture prey as it emerges. They are also capable of moving over semi-flooded mud-flats because they still possess over-lapping ventral scales. Although adults are usually found below the low tide, juveniles are intertidal in habit.

The three mangrove seasnakes may be among the most primitive seasnakes known. However, their natural history, biology and venom toxicity is poorly documented, primarily because only a handful of specimens have been found.

Range: Northern Australia and the Aru Islands, Indonesia • Max. length: 0.5m • Venom: Presumably neurotoxic and myotoxic, poorly known; no snakebites recorded • Habitat: Coastal mangrove mud-flats and estuaries • Prey: Gobies, mudskippers and small crustaceans • Reproduction: Presumably viviparous; a female with three embryos is recorded.

Similar species: Grey's mangrove seasnake (*Ephalophis greyi*) and Darwin mangrove seasnake (*Hydrelaps darwiniensis*) are also found in northern Australian coastal mangrove swamps.

Pelagic seasnake *Pelamis platurus*

This snake is the most distinctive of all seasnakes, patterned with yellow and black as a warning of its deadly capabilities. From the Persian Gulf it ranges southwards to South Africa, just entering the Atlantic Ocean near Cape Town and with occasional records from Namibia. East of the Gulf, its range includes and exceeds those of all other seasnakes. To the north it is recorded from Japan, Korea and Siberia, to the south it reaches Perth, Tasmania, and even New Zealand. Across the Pacific there are records from Micronesia, Polynesia, Hawaii, the Galapagos and the Pacific coast of the Americas from California to Ecuador. The South African, Japanese and New Zealand records are probably waifs, carried by the currents, but this is still an incredible distribution for a reptile.

The pelagic seasnake can live far from land, drifting with the currents or swimming strongly, forwards or backwards. It can also dive to 50m and remain submerged for over three hours. But it is helpless on land and individuals beached by storms will die. Although an open water species it has a preference for shallow waters in the vicinity of land-masses. Thousands-strong aggregation of pelagic seasnakes, may drift in slicks of floating debris along the convergence zones between two currents in bays and river mouths. This seasnake feeds on small fish sheltering under the floating debris or its own coils. It has few natural predators – even sharks avoid it – and vultures scavenging on the beach refuse dead pelagic seasnakes. Naive predatory fish from the Atlantic, experimentally introduced to the snake, will attack and consume it, often dying from its bite.

Range: Indian and Pacific Oceans • Max. length: 1.1m • Venom: Postsynaptic neurotoxins and myotoxins; human fatalities recorded • Habitat: Open water of bays and river mouths • Prey: Pelagic fish generalist • Reproduction: Viviparous; 2–6 neonates.

Left: *The Pelagic seasnake (*Pelamis platurus*) is the most widely distributed, naturally occuring snake in the world being found across two great oceans from South Africa to North America, the long way around.*

Below: *The Arafura mangove seasnake (*Parahydrophis mertoni*) is one of three rare seasnakes which inhabit the mangrove swamps of northern Australia and eastern Indonesia.*

Global Distribution of Venomous Snakes

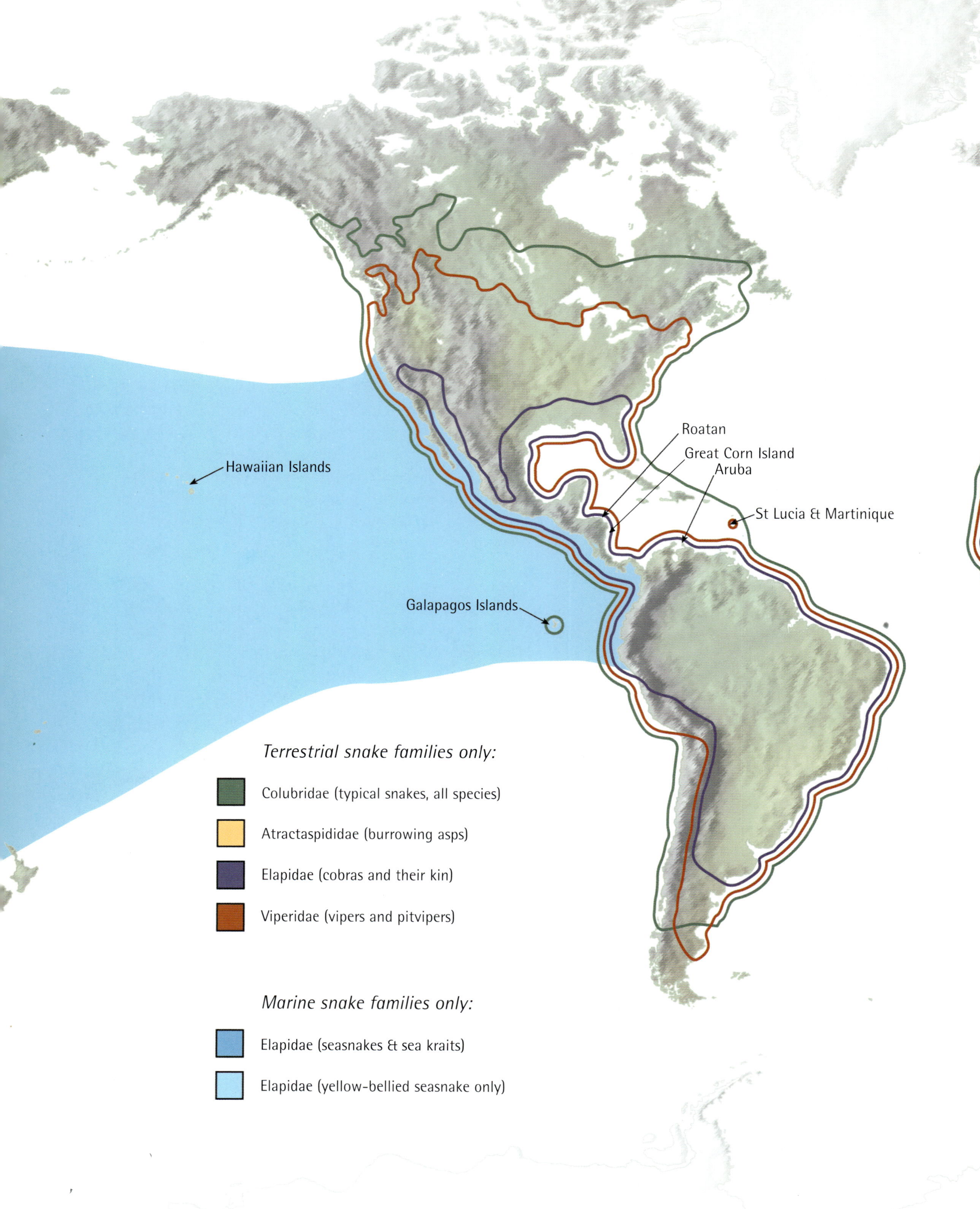

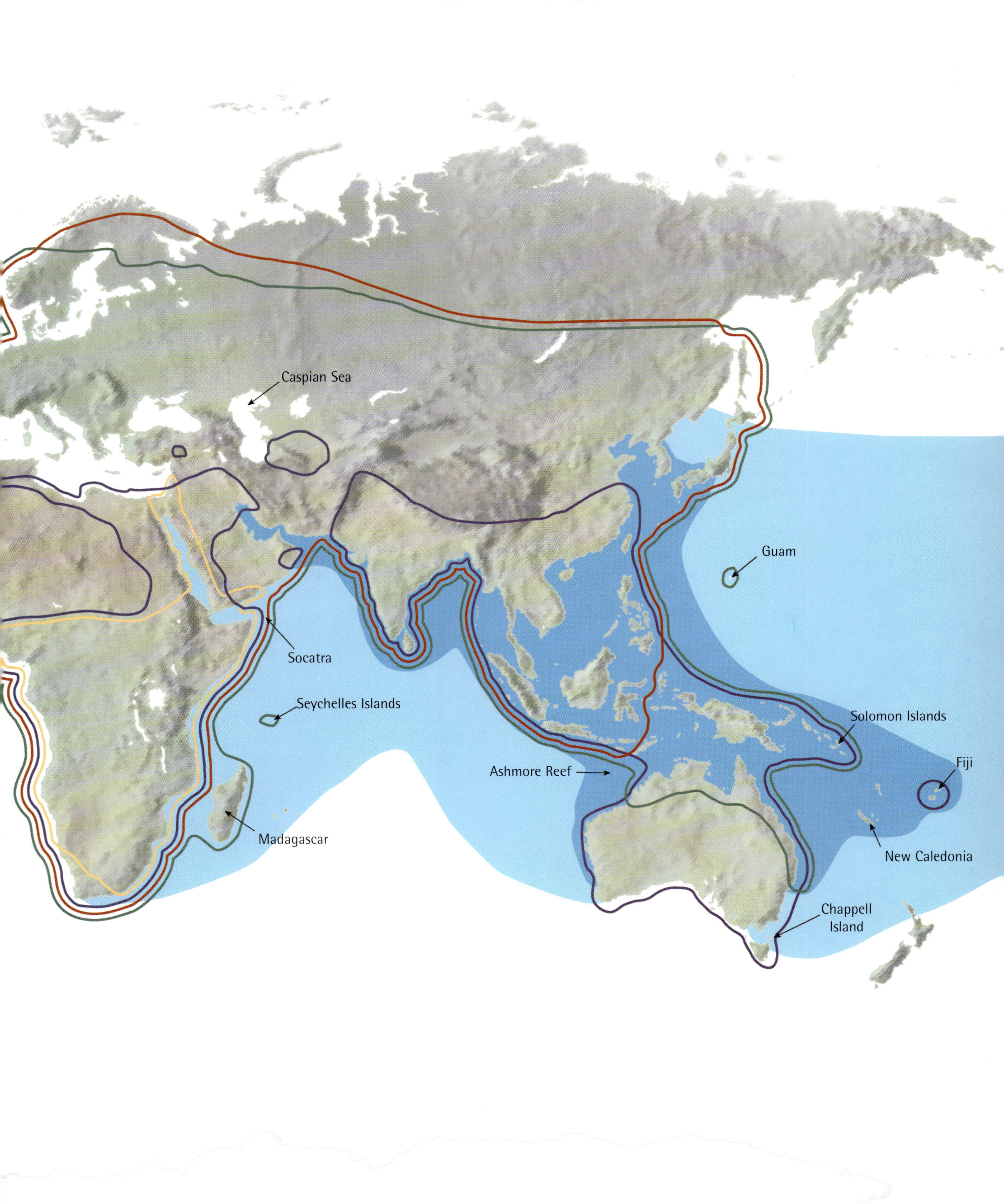

Caspian Sea
Guam
Socatra
Seychelles Islands
Solomon Islands
Fiji
Ashmore Reef
Madagascar
New Caledonia
Chappell
Island

Antivenom

The only specific antidote for the treatment of life-threatening snakebites is antivenom.

Antivenom is usually produced by immunizing horses or sheep with small but increasing doses of snake venom to stimulate their immune systems to generate antibodies that will neutralize the venom components. At intervals the animals are bled, the plasma is separated and the red blood corpuscles returned to them. The plasma containing the antibodies may be put through a number of filtration and purification processes until an antivenom (also known as antivenin, anti-snake venom serum, or just antiserum) is produced that can be given intravenously to human snake bite victims. Ideally, antivenoms should be kept in the dark between 2 and 8 degrees centigrade but even the less stable liquid preparations will retain most of their activity for more than a month in tropical temperatures, while freeze-dried antivenoms are stable for longer periods. They are therefore more suitable for the many remote parts of the world where refrigeration is a problem.

Antivenom, may be monovalent, where the horse has been immunized against the venom of a single species of snake or polyvalent where the horse has been immunized against the venoms of several snakes (both elapids and viperids) of greatest medical importance in a particular geographical region. Generally, polyvalent antivenoms must be administered in larger quantities to have any effect, since they contain antibodies to several different venoms. As a result of this higher dose, unpleasant or dangerous serum reactions may be more frequent. However, where the victim has not seen the snake that bit them, or the doctor is unsure of its identification from the patient's symptoms, the polyvalent antivenom is the safer bet. Monovalent antivenoms are produced for African species like mambas and the boomslang.

There is no doubt antivenom has saved many lives since its introduction. Before monovalent death adder antivenom was produced in Australia in the 1950s, some 50 per cent of serious death adder bites were fatal but today nobody should die from their bites.

Today, upwards of 40,000 people probably still die of snakebite annually, mostly in the tropical third world and especially in Sri Lanka, India, Nepal, Myanmar, New Guinea and Nigeria. Even more snake-bitten people are left with permanent physical handicaps following amputation of gangrenous extremities. There are still bites by some snakes for which there is no antivenom, such as some Asian kraits, the mangrove pitviper, burrowing asps and the African twigsnake.

Some antivenoms are extremely expensive. The new North American rattlesnake antivenom (CroFAb) costs more than ten times the price of the previous, admittedly not entirely satisfactory product (Wyeth). In the 1990s, Papua New Guinea was spending 50 per cent of its drug budget (0.5 per cent of its entire health budget), on imported Australian antivenom.

Antivenom is not only becoming unaffordable, it is becoming totally unobtainable. Animal rights protests caused the closure of three Indian antivenom production companies and stocks are dwindling in a country that already suffers tens of thousands of snakebite fatalities a year. The European pharmaceutical companies that produce antivenoms for Africa or Asia are either limiting their production to antivenoms for North America and Europe (Protherics), or they are pulling out of antivenom production all together (Behringwerke), mainly because there is little or no money in producing drugs specifically for third world countries that cannot afford to pay full market price.

However, excellent antivenoms are still being produced in Thailand, South Africa, Costa Rica and Brazil. Producers in these countries are being persuaded to make antivenoms for neglected areas such as West Africa. Two new 'Pan African' polyvalent antivenoms are now ready for clinical trials in Nigeria.

Further Reading

It is not feasible to list every national field guide and every historical scientific paper covering venomous snakes. Therefore the list below is confined to fairly recent and available regional or general publications on the subject which will guide the reader towards more specific national, generic or species literature sources.

General and Introduction:

Gloyd H. K. and R. Conant (1990) Snakes of the *Agkistrodon* Complex: A Monographic Review. *Society for the Study of Amphibians and Reptiles.*

Greene H. W. (1997) *Snakes: The Evolution of Mystery in Nature*. California Univ. Press.

Mallow D., D. Ludwig and G. Nilson (2003) *True Vipers: Natural History and Toxinology of Old World Vipers*. Krieger Publishing Company, Malabar, Florida.

Meier J. and J.White (eds.) (1995) *Handbook of Clinical Toxicology of Animal Venoms and Poisons*. CRC Press, Boca Raton.

Phelps T. *at press*. Old World Vipers: Natural History of the Azemiopinae, Causinae and Viperinae. *Chimaira*, Frankfurt am Main.

Pough F. H., R. M. Andrews, J. E. Cadle, M. L. Crump, A. H. Savitzky and K. D.Wells (2004) *Herpetology* (3rd edition). Pearson Prentice Hall, New Jersey.

Rodda G. H., Y.Sawai, D. Chiszar and H. Tanaka (eds.) 1999 *Problem Snake Management: The Habu and the Brown Treesnake*. Comstock Cornell.

Schuett G. W., M. Höggren, M. E. Douglas and H. W. Greene (eds.) (2002) *Biology of the Vipers*. Eagle Mountain Publishing, Utah.

Thorpe R. S., W. Wuster and A. Malhotra (eds.) (1997) *Venomous Snakes: Ecology, Evolution and Snakebite*. Zoological Society of London.

Zug G. R., L. J. Vitt and J. P. Caldwell (2001) *Herpetology: An Introductory Biology of Amphibians and Reptiles*. (2nd edition). Academic Press.

Chapter One – Americas:

Campbell J. A. and W. W. Lamar (2004) *The Venomous Reptiles of the Western Hemisphere* (Volumes I and II). Comstock-Cornell Publishers, Ithaca.

Klauber L. M. 1997 *Rattlesnakes* (Volumes I and II) (2nd edition). Univ. California, Berkeley.

Mattison C. 1996 *Rattler! A Natural History of Rattlesnakes*. Blandford Press.

Roze J. A. (1996) *Coral Snakes of the Americas*. Krieger Publishing Company, Malabar, Florida.

Rubio M. 1998 *Rattlesnake: Portrait of a Predator*. Smithsonian Inst. Press.

Chapter Two -Eurasia:

Arnold E. N. and D. W. Ovenden (2002) *Reptiles and Amphibians of Europe* (2nd edition). Princeton University Press.

Chapter Three – Africa:

Broadley D. G. (1983) *FitzSimons' Snakes of Southern Africa*. Delta Books, Johannesburg.

Campbell J. A. and E. D.Brodie Jr. (eds.) (1992) *Biology of the Pitvipers*. Selva Publishing, Tyler, Texas.

Spawls S. and W. Branch (1995) T*he Dangerous Snakes of Africa*. Blandford Press.

Spawls S., K. Howell, R. Drewes and J. Ashe (2002) *A Field Guide to the Reptiles of East Africa*. Natural World, Cape Town.

Visser J. and D. S. Chapman (1978) *Snakes and Snakebite: Venomous Snakes and Management of Snakebite in Southern Africa*. Purnell, Cape Town and Johannesburg.

Chapter Four - Tropical Asia:

Gumprecht A., F. Tillack, N. L. Orlov, A. Captain and S. Ryabov (2004) *Asian Pitvipers*. GeitjeBooks, Berlin.

Chapter Five – Australasia:

Cogger H. G. (2000) *Reptiles and Amphibians of Australia* (6th edition). New Holland.

Mirtschin *P. and R. Davis (1982)* Dangerous Snakes of Australia. Rigby Publishing, Adelaide.

O'Shea M. (1996) *A Guide to Snakes of Papua New Guinea*. Independent Publishing, Port Moresby.

Wilson S. and G. Swan (2003) *A Complete Guide to Reptiles of Australia*. New Holland.

Chapter Six – The Oceans:

Dunson W. A. (ed.) (1975) *The Biology of Sea Snakes*. Univerity Park Press, Baltimore.

Heatwole H. (1999) *Sea Snakes* (2nd edition). Krieger Publishing Company, Malabar, Florida.

The papers cited in the Phylogeny of Snakes table on page 18:

Vidal (2002) Colubrid systematics: Evidence for an early appearance of the venom apparatus followed by extensive evolutionary tinkering. *Toxinology of Colubrid Snakes: Biology, Venoms and Envenomation Journal of Toxicology* 21(1+2):21-41.

Vidal and Hedges (2002) Higher-level relationships of snakes inferred from four nuclear and mitochondrial genes. *C. R. Biologies* 325:977-985.

Fry B. G. and W. Wuster (2004) Assembling an arsenal: Origin and evolution of snake venom proteome referred from phylogenetic analysis of toxin sequences. *Molecular Phylogenetic Evolution*. 21(5):870-883. also Fry, B. G. N.

G.Lumsden, W. Wuster, J. C. Wickramaratna, W. C. Hodgson & R. M. Kini (2003) Isolation of a neurotoxin (a-colubritoxin) frm a nonvenomous colubrid: Evidence for early origin of venom in snakes. *Journal of Molecular Evolution* 57:446-552.

Index

Page numbers in *italics* denote illustrations; names in **bold** denote featured species.

Picture Credits

© Mark O'Shea: 2, 10, 12, 13, 15 (t, cl, cr, bl, br), 24, 25, 26, 27 (l, r), 31 (t, b), 32 (t, b), 33 (b), 35 (t, b), 36, 38, 39, 40 (b), 41 (t), 46 (r), 47, 49 (t), 51 (t, b), 52 (b), 58 (t, b), 60, 64 (c), 67, 70 (l, r), 71 (t), 73 (l), 75, 76 (b), 80 (t, b), 81 (b), 82 (t, b), 85 (l), 87, 88 (t, b), 89 (b), 90, 94, 95, 96, 97, 98, 100, 103, 105 (t, b), 108, 110 (b), 114 (l, r), 115 (b), 116, 118, 119, 120, 121, 123 (b), 125, 126, 130 (t, b), 132, 133, 134, 140, 141, 142, 143, 145, 146, 147, 148

© Daniel Heuclin/NHPA: 4 (*see also* 107), 28, 34, 37, 40, 46 (l), 48, 56, 57 (t, b), 59, 62 (t, b), 63, 64 (b), 68, 73 (r), 74 (br), 81 (t), 84 (l), 93, 99, 101, 124, 137, 138

© James Carmichael Jr/NHPA: Front cover, 8, 33 (t), 61 (t, b) 77 (b), 85 (r), 109

© David A. Warrell, 20, 21, 22 (l, r), 52 (t), 66

© Jany Sauvanet/NHPA: 30, 53

© Wolfgang Wuster 41 (b)

© Bill Love/NHPA: 42, 43, 45, 83, 106, 111

© Stephen Krasemann/NHPA: 44

© Kevin Schafer/NHPA: 50

© Joe Blossom/NHPA: 54

© John Hartley/NHPA: 64 (t)

© Nigel J Dennis/NHPA: 71 (b)

© Anthony Barrister/NHPA: 72 (t, b), 76 (t), 77 (t), 78, 79, 86 (c, b), 89 (t)

© Steve Spawls: 74 (bl)

© Martin Harvey/NHPA: 84 (r), 86 (t)

© Hellio & Van Ingen/NHPA: 92

© Nikolai Orlov: 102, 104, 110 (t)

© Ken Griffiths/NHPA: 128, 131 (b)

© ANT Photo Library/NHPA: 112, 115 (t), 117, 122 (b), 129, 131 (t) 136 (t, b), 144, 150

© Cliff & Dawn Frith/ANTPhoto.com: 122 (t)

© Ralph & Daphne Keller/NHPA: 123 (t)

© Kark Switak/NHPA: 49 (b), 127

© Ian McCann/ANTPhoto.com: 129

© Rupert Russell/ANTPhoto.com: 131 (t)

© Mike McCoy: 135 (t, b), 136 (b)

© Adrian Hepworth/NHPA: 151

Acknowledgements

This book would not be as comprehensive without the assistance of friends and colleagues Bill Love, Nikolai Orlov, Mike McCoy, Steve Spawls and Wolfgang Wuster, who provided images and expert opinion at short notice. Advice and constructive criticism of the manuscript by Professor David A. Warrell, who also provided images of rare snakes and the results of snakebite, was particularly helpful. Many thanks to my partner, Bina Mistry, for assisting with proof-reading, acting as a non-specialist commentator and sharing me with the word processor. Thanks also to Charlotte, my long-suffering editor who was forced to come to terms with many tongue-twisting scientific names.